Understanding Children's Mathematical Graphics

Understanding Children's Mathematical Graphics

BEGINNINGS IN PLAY

Elizabeth Carruthers and Maulfry Worthington

Open University Press

Open University Press
McGraw-Hill Education
McGraw-Hill House
Shoppenhangers Road
Maidenhead
Berkshire
England
SL6 2QL

email: enquiries@openup.co.uk
world wide web: www.openup.co.uk

and Two Penn Plaza, New York, NY 10121-2289, USA

First published 2011

A catalogue record of this book is available from the British Library

ISBN-13: 978-0-33-523776-0 (pb) 978-0-33-523775-3 (hb)
ISBN-10: 0-33-523776-2 (pb) 0-33-523775-4 (hb)
eISBN: 978-0-33-524079-1

Library of Congress Cataloging-in-Publication Data
CIP data applied for

Typeset by RefineCatch Limited, Bungay, Suffolk
Printed an bound by CPI Group (UK) Ltd, Croydon

The **McGraw·Hill** Companies

To my husband Tom and to all my childhood friends in Galston, Ayrshire, Scotland.

Elizabeth

And for many happy memories of a childhood full of play and sunshine in Hampshire and Dorset.

Maulfry

Contents

SECTION 2

Pedagogy and practice

List of figures

List of case studies

About the authors

Maulfry and Elizabeth have both taught in the full 3–8 year age range for over 26 years. They have conducted extensive research in key aspects of early mathematics and have many publications including *Children's Mathematics: Making Marks, Making Meaning* (Sage Publications 2006). They are winners of several national awards for their work with children and teachers on *children's mathematical graphics*. Their work is featured in the *Williams Maths Review* (DCSF 2008) and they were commissioned to write *Children Thinking Mathematically* (DCSF 2009).

Maulfry and Elizabeth are founders of the international 'Children's Mathematics Network', a grassroots organization for teachers and educators, supporting research and pedagogy for children's mathematics in the birth–8 year age range: www.childrens-mathematics.net Their innovative professional development initiative for Early Years mathematics is featured in the final report of 'Researching Effective CPD in Mathematics Education' (RECME), (NCETM 2009).

Elizabeth is presently headteacher of Redcliffe Children's centre and Maintained Nursery School, Bristol. She has worked in local authorities as an Early Years Advisor and as a National Numeracy Consultant. She is currently researching the pedagogy of *children's mathematical graphics*, children's wild outdoor experiences and leadership in Children's Centres, and is a doctoral student at the University of the West of England in Bristol.

Maulfry has also worked as a National Numeracy Consultant and has lectured on Early Years education, and Primary and Early Years mathematics. She is currently engaged in doctoral research at the VU University, Amsterdam, conducting a longitudinal, ethnographic study into the emergence of *children's mathematical graphics* in imagination and symbolic play, with children of 3–4 years of age.

Foreword
Janet Moyles

Each picture told a story; mysterious often to my undeveloped understanding . . .
yet ever profoundly interesting.

(Brontë 1994: 10–11)

When asked to write this foreword, I confess to having a very 'undeveloped understanding' of *children's mathematical graphics* as determined by the book title. While children's pictures and drawings have always held a profound interest for me and many other early years educators, the rationale behind children's representations and creations has often eluded me and I have resorted, no doubt like many readers, to saying, 'Tell me about your picture,' in the hope of some elucidation of what I might do or say next to the child! Within a few pages of starting to read this book, I felt a certain 'guilt' that I could have done considerably more to understand the meanings behind children's playful representations, especially in mathematics.

Graphics appear to be so much more than 'drawing' – graphics are ways of representing thinking, in this case, that of children from birth to 6 years. As Embree (2010) enquires: 'Why do we include graphics . . . to summarize computations, to explain difficult ideas . . . A graphic demands at least as much care as the text that surrounds it, often rather more'.

This is exactly what children are trying to do, according to the authors of this book – explain their own ideas and computations. Through a series of carefully observed and convincing examples of children's play, the authors outline their understanding of how *children's mathematical graphics* develop. They show how, with sensitive adult interaction, analysis and interpretation, what might be described as playful 'scribbles' or 'marks' made by children can have a deep, well-conceived meaning to the individual child and relate closely to his or her intellectual development.

The authors make it clear that 'graphics' is so much more than merely an 'outcome'. Graphics in mathematical experiences are part of the processes of maths learning and the whole problem-generating and problem-solving ethos that we try, as educators, to develop in, and for, our learners. Those playful mathematical experiences and challenges which are initiated by the child are likely to be richer and more motivating as well as being a stimulus for children's sense of self-worth, self-confidence and competence. They are also more likely to ensure that the children's voices are clearly heard among the (necessary but sometimes overwhelming) bureaucracy and direction which seem to permeate curriculum and assessment, in England especially. 'Child-initiated experiences rely on the child being competent and knowledgeable about their own needs and choices' (Moyles 2008: 33), a point also stressed throughout the chapters of this book.

The emphasis we, as educators, tend to put on 'outcomes' of maths and other 'work' for children can militate against them developing their own deeper cognition and indicate that adults are sometimes working against the children's understanding rather than extending it. Practice, as explained in this book, should be geared towards adults' waiting, really listening to children, and, in so doing, thoroughly understanding what children can already achieve in mathematical thinking through their representations, leading to knowledge of how to extend their learning. Superb examples proliferate in this book which will effectively support playful pedagogies (Moyles 2010).

The importance of appropriate pedagogies and the building of environments that support children's mathematical thinking are key factors in determining success for educators and children. The adults' roles – and their reflection on those roles and provision for learning (Moyles 2010) – cannot be overstated, especially in observing and planning for spontaneous mathematical opportunities for learning, both of which are richly expressed in Elizabeth Carruthers' and Maulfry Worthington's writing. As Goouch (2010: 55–6) asserts: 'To accompany children in their play is a sophisticated role that can be achieved only by those who know and understand children, who are able to allow the sometimes complex intentions of children at play to take precedence and who will demonstrate respect for such intentionality'.

The emphasis on the 'intentionality' of young children's graphics is an excellent point for readers to appreciate, given that children's actions sometimes defy adult logic! Children are not merely making marks but more often than not have logical intentions behind the shapes, forms and lines they represent. It is up to the adults to interpret, alongside the children, what are their deeper intentions and meanings, sharing power and knowledge and achieving success and motivation in mathematics.

The close links between *children's mathematical graphics*, algebra and trigonometry are undisputable and while it may seem that young children are still a world away from needing that level of mathematical understanding, it is clear in this book that they start at a very early age instinctively to develop these complex ideas. Aside from children, the book will really make readers explore the concept of mathematics and graphics as a personal understanding of a complex system which children – if freed from directive teaching – can readily use to make sense of maths for themselves.

One cannot fail to be enamoured throughout this book by the thoroughness of the arguments and examples put forward by the authors as well as the way in which the writing is totally embedded in children and their daily graphical/mathematical experiences. The taxonomy of *mathematical graphics* developed by the authors over several years from many meticulous observations will effectively enable early years educators to sustain children's mathematical learning and understand for themselves how development and progress occurs. If you struggle to teach maths and are concerned about 'written' evidence, then you will appreciate the comprehensive coverage of the issues surrounding maths teaching to young children, especially the birth-to-3s. Even those who are already keen and skilled early years maths teachers will enhance their own learning and practice substantially through absorbing the contents of this book.

Returning to the beginning and the Charlotte Brontë quote, the many examples of children's graphical representations contained in this excellent book have certainly proved 'profoundly interesting', have unravelled the 'mysterious' and enabled me to develop my very 'undeveloped understanding' of this area. The many illustrations of *children's mathematical graphics* tell their own story, as is illustrated in the oft-quoted saying: 'A picture [in this case a *graphic*] is worth a thousand words'. As very young children often cannot express what they mean in words, our recourse, as educators, must be towards their graphics and other representations.

I feel certain that, through this book, readers will develop their mathematical knowledge and that young children's intellectual and emotional development will be enhanced as a result of early educators using the taxonomy and carefully conceived ideas generated.

References

Brontë, C. (1994), *Jane Eyre*. Glasgow: Penguin.

Embree, M. (2010) *The Design of Mathematical Graphics Theory and Practice*, www.caam.rice.edu/~siamchapter/files/embree_graphics.pdf, accessed 30 July 2010.

Goouch, K. (2010) Permission to play, in J. Moyles (ed.) *The Excellence of Play*, 3rd edn. Maidenhead: Open University Press.

Moyles, J. (2008) Empowering children and adults: play and child-initiated learning, in S. Featherstone and P. Featherstone (eds) *Like Bees Not Butterflies: Child Initiated Learning in the Early Years*. London: A&C Black.

Moyles, J. (ed.) (2010) *Thinking About Play: Developing a Reflective Approach*. Maidenhead: Open University Press.

Preface

We wrote this book firstly to illustrate and explain the many ways children use their *mathematical graphics* in their play. Secondly, there are extremely few publications that explore play and its relationship with early 'written' maths.

We felt there was an overwhelming need to highlight the case for real play and its relationship to children's mathematics. Real play belongs to children; they choose the resources, the place and the children they play with. They make, explore and communicate their personal meanings through imagination and play. Teachers who really understand the children's perspectives and who deeply observe and note children's meaning-making feel they are learning more and more from the children: these are the teachers who are on the 'inside' of children's learning. Vivian Gussin Paley, a teacher for more than 50 years, is a true insider, and for this reason we have quoted her observations of children. She argues that play is 'the curriculum the children have for themselves' (Paley 2004: 20).

The first section of this book focuses on children's meaning-making through their imagination, symbolic play and graphicacy (i.e. drawing, maps and writing), with numerous case studies of *children's mathematical graphics* in play. The second section explores pedagogical issues and includes a range of leadership voices from those leading and learning about *children's mathematical graphics*.

Children's mathematical graphics are totally inclusive. The wonderful examples in this book reveal children's *own* mathematics and show how powerful young children's thinking is from birth to 7 years. We believe that every child is exceptional: to borrow from Papert, 'The children in these studies are not exceptional, or rather, they are exceptional in every conceivable way' (1993: 13). With sensitive teachers and practitioners who take an interest and support children's meaning-making, all young children can reach out and grasp their mathematical potential: all children can fly.

Acknowledgements

Above all, our warm thanks to all the children for allowing us to share in your imaginations . . .

We should like to thank the staff at the following early years settings and schools who have so generously shared their observations and examples of children's imagination and symbolic play, their graphicacy and *children's mathematical graphics*.

Our special thanks to all the staff at Redcliffe Children's Centre, Bristol; especially Emma Butcher, Carole Keane and Hugo Turvey, who are true insiders of children's learning. Our thanks also to Stephanie Ager, Lisa Allen, Sue Cook, Susan Dyer, Louise Evans, Jeanette Hill and Clare Reed and Maria Bouyamourn, who have contributed to this book. Thank you also to Donna Andrews, Amanda Clancy, Philippa Cook, Biddy Jones, Beth Osborn and Julia Sutcliffe of the Foundation Stage, Clifton High School, Bristol, for sharing their enthusiasm for *children's mathematical graphics*, and for their generous contributions of so many examples.

Grateful thanks to Juliette Aubrey and Ann Foulkes at Little Folks Nursery School, Limpley Stoke, Bath; Karen Eggenschwiler and Emma Hookey at the International School, Kilchberg, Zurich; Milena Annecchiarico and Sarah Ryan at Barnsole Infants School, Gillingham, Medway; Lesley Pease: Early Years Advisor, Medway, for her continuing interest and generous support. Thanks to Donna Hackney, Jo Shepherd and Keeley Woolley at Rainbows Pre-school, Rainham; Bev Howland, Borbala Unger and Jan Wadlow at the Foundation Stage Unit at Pilgrim School, Rochester; Jan Butler at St John's Infants CE primary school, Chatham; and Maxine Watson at Byron Primary School, Gillingham, Medway.

Our appreciation and thanks to the teachers in the Netherlands: Jo-nele Blokzijl, Katja de Vries; Francisca Kramer and Marieke Ploeger at the Julianaschool, Schagen, and to Maartije Roodnat of the Basischool de Avonturijn (Balibar nursery class) in Amsterdam for sharing your interest and pedagogy.

Thanks also to Reception teacher Kylie McBeth at Princess Frederica CE VA Primary School, Kensal Green, Brent, London, and to the following teachers from Bristol: Sara Keirle at St Peter's CE School; Janet Martin, Avonmouth CE Primary School; Becky Perry, Filton Avenue Infants School; Karen Sheppard, Blaise Primary and Nursery School. Thank you Samantha Mosely at Freshford Primary School, Bath, and Kathryn Oram at Victoria Park Infants School, Bristol. Thanks to Jane Brough, Sally Hesbrook and Donna Williams and at St Mary Redcliffe Infants School, Bristol. From our 'Project 2003', our appreciation also goes to Ros Arminshaw and Michelle Barrett of the Rowland Hill Centre for Childhood, Haringey; Louise Glovers at the Robert Owen Children's Centre, Greenwich, London, and Julie Mills from Bramble Hedge Pre-school, Plymouth.

For contributions to the text our thanks go to Lesley Pease, Early Years Advisor, Medway, Kent; Joy Harper, African and African-Caribbean Achievement Consultant, Brent local authority; Emma Hookey and Karen Eggenschwiler from Zurich and Emma Higgins and Carole Keane at Redcliffe Children's Centre and Maintained Nursery, Bristol.

Last – but not least – our special thanks go to Fiona Richman and Stephanie Frosch of Open University Press for their sympathetic support, and for helping us realize this book.

Permissions acknowledgements

Figures 2.11; 2.12; 2.13; 2.14; 2.15; 2.16; 4.11; 4.14; and 4.15 were first published in our book: Carruthers, E. and Worthington, M. (2003) *Children's Mathematics, Making Marks, Making Meaning*. London: Sage Publications. Figure 4.11 first appeared on the cover of the 2nd edition (2006). Reproduced with kind permission of Sage Publications.

Figures 3.11; 4.2; 4.3; 4.4; 4.8; 4.9; 4.10 and 4.12 first appeared in a paper by Worthington, M. (2009) 'Fish in the water of culture: signs and symbols in young children's drawing', *Psychology of Education Review* (PER), Volume 33, Number 1, March, 37–46, and are included here with kind permission of the editor of the *Psychology of Education Review*.

Figures 6.1 and 6.2 were first included in the article Carruthers, E. and Worthington, M. (2009) 'Children's mathematical graphics: understanding the key concept' *Primary Mathematics*, Autumn, Volume 13, Issue 3, 3–5. They are reproduced here with kind permission of the editor of *Primary Mathematics*.

Figures 7.26 and 10.1 were first published in Carruthers, E. and Worthington, M. (2009) 'Learning & development: mathematics – marking time', *Nursery World*, 30 September, 24–25, and are included here with kind permission of the editor of *Nursery World*.

Figures 2.3; 2.4; 2.9; 2.10 and 7.27 first appeared in Worthington, M. (2010) 'Play is a complex landscape: imagination and symbolic meanings' in P. Broadhead, L. Wood and J. Howard (eds) *Play and Learning in Educational Settings*, London: Sage Publications, and are reproduced here with kind permission of Sage Publications.

Figure 2.7 was first published in Worthington, M. (2010) '"This is a *different* calculator – with computer games on": reflecting on children's symbolic play in the digital age', in J. Moyles (ed.) *Thinking About Play: Developing a Reflective Approach*.

Maidenhead: Open University Press. It is included here with kind permission of Open University Press.

Figures 4.13; 7.14 and 7.15 first appeared in 2011: Worthington, M. (2011) 'Young mathematicians: global learners' in E. Murphy (ed.) *Welcoming Linguistic and Cultural Diversity in Early Childhood Classrooms*. Bristol: Multilingual Matters. They are reproduced here with kind permission of the editor.

SECTION 1
Children

SECTION 2

1

Play and mathematics

‘ *"Why is it," Milo asks, "that quite often even the things which are correct don't seem to be right?" Why indeed? I have hideous memories of taking O level Maths. I had at last, triumphantly got one problem solved. The answer was in sheep. Then I looked at the end of the question. It said, bafflingly, "Give the answer in bags of coal." It was like a bad dream.'*

(Juster 2008: 4)

What is this chapter about?

- **The importance of play** and the relationship between play and mathematics in early childhood.
- **Making and communicating meanings** – some alternative symbolic systems in other cultures; an introduction to social semiotics and multimodality; communicating mathematical meanings.
- **Why maths matters** – and some of the challenges concerning teaching early 'written' mathematics.
- **A new perspective of play** that supports children's understanding of the abstract symbolic 'written' language of mathematics.

This book focuses on young children from birth to 6 years in educational settings as they explore, make and communicate meanings. It provides both context and theoretical underpinnings supported by numerous examples and case studies. Above all, the book celebrates young children's imagination and play and their ability to use their own graphical marks and symbols to make and communicate their mathematical thinking.

The importance of play

This book is both play and maths 'rich'. It interweaves children's graphicacy throughout (i.e. drawing, maps, writing and *children's mathematical graphics*), revealing the holistic nature of the young child's learning and their amazing interests and abilities in this significant period of their lives.

The book challenges traditional beliefs and practices of teaching 'written mathematics' in early childhood, to 'set maths free' and empower young children. Empowerment has been shown to increase children's confidence in mathematics and in turn to support their understandings of the abstract, symbolic written language of mathematics in ways that make personal sense.

Our work is based on poststructural and democratic perspectives that privilege young children's power, perspectives, meanings and 'voice'. We hope that readers will also be set free of any previous anxieties they may have had about teaching mathematics, and see that, by truly valuing and supporting children's meaning-making and graphicacy *every* young child can achieve in mathematics. And, since almost all of the examples originated in the children's free, self-initiated play, we believe that this book offers some very important and positive messages about how incredibly rich with potential the combination of imagination and symbolic play can be.

Play is highly complex and for adults it can often appear confusing and difficult to understand. Teachers' uncertainty over play may be partly due to the fact that there is no one definition of the term 'play'. Genuine child-initiated play is spontaneous and belongs to the child: it offers a 'non-threatening way to cope with new learning and still retain self-esteem and self-image' (Moyles 1994: 7) and has a 'free-flow' quality (Bruce, 2005). Contemporary research on play is informed by cultural-historical theories and guides early childhood policies and curricula in western societies (Wood 2009). According to Vygotsky (1978), one of the features of effective play is that it is social.

Wood argues that 'what play is, what play means on and what play does for the players is conceptualised in different ways according to the particular lenses through which researchers view play' (2010: 12). Moyles points out that the findings from recent research on play emphasize 'children's voices and children's choices . . . the research illustrating children's intentions, motivations, meaning and modes of engagement as well as exemplifying the adults' roles' (2010: xiii), revealing children's learning and development from a range of perspectives. 'Play' can only truly be described as play when it is the child's own, and no amount of 'adult-planned play' deserves to be described in this way. There is a distinct difference between children's free and self-initiated play (i.e. play that belongs to children, that arises from their own interests and cultural experiences and empowers them to follow their own enquiries) and 'playful activities', which are those that belong to the adults and often have set outcomes.

There is a long history of research in this country and internationally (e.g. Isaacs 1929; Sylva *et al.* 1980; Pellegrini and Smith 2005; Athey 2007) that explores children's play from many perspectives. Additionally, play has been the subject of play scholars such as Huizinga (1950), Sutton-Smith (1997) and Caillois (2001), while Goswami (2008) has shown play to be significant for children's cognitive development. However, Broadhead (2004, cited in Broadhead *et al.* 2010: 181) argues that in England our 'play heritage was substantially eroded by the culture and climate of educational reform from the late 1980s onwards . . . to the extent that the status of play in education has been at its lowest point in the last 20 years.'

Play and mathematics

A lack of child-initiated play severely limits possibilities for children to explore and communicate their own interests and mathematical ideas. It also restricts opportunities for children to engage in the sort of dialogue that can scaffold their understanding about their graphical marks and symbols, limiting their mathematical thinking and communication. This will result in imbalanced and impoverished experiences and has implications for their understanding of 'written' mathematics.

Rich child-initiated play supports children's mathematical thinking and can be considered in three (non-hierarchical) ways; all are important.

- **Symbolic play** underpinning graphicacy and the abstract, written language of mathematics.
- **Exploring and communicating mathematical thinking** through graphical representations in various play contexts.
- **'Practical' mathematical play**.

In their *symbolic play* children build on their earliest awareness of relationships between objects, signs and meanings (the symbol-meaning-communication relationship). Children's meaning-making in imagination and symbolic play underpins *children's mathematical graphics*.

By *exploring and communicating mathematical thinking* through graphical representations, children follow their own interests and develop 'meaning-full' *mathematical graphics*: these mathematical explorations are *not* adult-planned or led.

In *practical mathematical play*, it is often the *resources* children use that are explicitly mathematical (e.g. number puzzles; tessellated shapes; skittles games; shop tills and money; calculators; balance scales, timers and rulers), or that offer mathematical possibilities (e.g. sand; water; collections of small artefacts). Since the mathematical potential of these resources is transparent, adults can readily identify the mathematics such as 'volume and capacity', 'sorting' or 'counting' in which the children are engaged.

This book focuses on the first two aspects we have described above, *symbolic play*, showing how it underpins *children's mathematical graphics*, and *exploring and communicating mathematical thinking* through graphical representations in various play contexts.

Making and communicating meanings

Vygotsky traced the beginnings of writing in gesture and play, arguing, 'Superficially, play bears little resemblance to the complex, mediated form of thought and volition it leads to. Only a profound internal analysis makes it possible to determine its course of change and its role in development' (1978: 104). Since both writing and mathematical notation are symbolic written languages, we argue that it is logical to assume that the beginnings of mathematical notation also have their origins in play (Worthington 2010a).

'Written' mathematics is of course not the entire mathematics curriculum, but it is a very significant aspect. Vile emphasises that 'There is an inextricable connection with signs and mathematics. One might even say that mathematics consists entirely of a complex system of signs . . .' (1999: 87), and van Oers describes mathematics as 'really a matter of problem solving with symbolic tools' (2001: 63).

Sand-talk and stick-talk

While 'written' mathematics involves pens, paper and other writing tools and surfaces, humans have not always communicated in this way and there are peoples today who continue to use other means and media, to communicate using symbolic languages that are specific to their cultures. These symbolic languages serve to emphasize that meaning-making and communication are universal human attributes.

For example, using sand as a medium, native Australian mothers in central Australia sometimes make marks and signs with their fingers in dry sand as they tell stories to their children or make maps. Chatwin (1987: 24) recounts how one Pintupu mother

> tells her tale in a patter of staccato bursts and, at the same time, traces the Ancestor's 'footprints' by running her first and second fingers, one after the other in a double dotted line along the ground . . . The sand drawings done for children are but sketches or 'open versions' of *real* drawings representing the real Ancestors . . . it is through the 'sketches' that the young learn to orient themselves to their land, its mythology and resources.

Detailing some of the conventions used in sand maps, Nash (1998) observes that sand-talk is used as 'a type of conversation', showing how such maps 'combine various meaning-making [semiotic] systems in the culture, notably the spoken language, gesture and iconography, as well as song, handsign and dance'.

Jenny Green has documented sand-talk among four native Australian tribes from a linguistics perspective, looking at, among other things, the different 'modes' of this language. She explains: 'A bunched, or a spread hand is used to make multiple individuated marks or dots with the tips of the fingers . . . the narrator is drawing fruits in a dish in this way' (see Figure 1.1). 'Rhythmic dotting made with the tips of the fingers is also used to represent particular types of motion, such as dancing' (2009: 138).

Descriptions and examples of sand-talk show that it shares some similarities with the gestural marks made by very young children when they begin to investigate the effect of their actions on the environment and in their early drawings (Matthews 1999). Sand-talk is an example of the world's 'transient languages', communicative systems that are temporary and impermanent. They include the system used by the Penan tribe of Borneo whose people, with no tradition of writing, communicate in 'a most extraordinary dialogue . . . by means of sign sticks, branches or saplings strategically placed and decorated with symbols that convey specific messages' in the rain forest (Wade *et al.* 1995: 52–3). 'Interpreting these symbols requires both knowledge of their individual meanings and understanding of the context in which the message was left' (1995: 53) – features that are significant for all symbolic languages.

Figure 1.1 Plural-marking hand shape
Source: Green, J. (2009) *Between the Earth and the Air: Multimodality in Arandic Sand Stories*, Ph.D. dissertation, Melbourne: University of Melbourne, page 138. Reproduced with permission.

While to our western perspective it may be difficult to appreciate sand-talk and sign sticks as symbolic languages, they are truly 'multimodal' communicative systems, providing a glimpse of the complexity and diversity of the languages of the world's cultures (see e.g. Agar n.d.). They emphasize the extent to which local knowledge is contained in different symbol systems and how these relate to peoples' lives, environments and specific cultural contexts (Harrison 2007). And, just as some indigenous Australians have communicated through sand-talk, and the Penan of Borneo use sticks to 'talk', young children create their own powerful ways of making and communicating meanings which we can identify, providing we are open to those ways.

Social semiotics and mathematics

Against this background there is a growing body of research into the role of representation in mathematics that is largely based on a Vygotskian, social semiotic perspective. In semiotics, 'representation' refers to a diverse range of symbolic tools and can include gesture, spoken words and artefacts as well as graphicacy, representations on paper or other surfaces. For there to be any *external* representations, there must also be *internal*, mental representations. In effect this brings the internal, mental representations 'out there': through their graphical representations, children's marks and symbols – and the meanings they embody – are mirrored back to them, revealing some of their thinking and promoting further reflections on the representations themselves.

The role of semiotics in mathematics generally has been well researched (e.g. diSessa *et al.* 1991; Cobb *et al.* 1992; Ernest 2006), although research has been limited in early childhood mathematics (e.g. Hughes 1986; Gifford 1990; Munn 1994; van Oers 2000). Pape and Tchoshanov propose that mathematical representations 'must be thought of as tools for cognitive activity rather than products or the end result of a task' (2001: 124).

Young children's *mathematical graphics* arise spontaneously though their perceived needs in play, contexts that offer realistic and personally meaningful situations in which to explore and communicate cultural knowledge and ideas. Later, when teachers and practitioners plan small group sessions they will want to ensure that the mathematics has a meaningful context and purposes that the children can understand. Vellom and Pape propose that 'typical' written mathematical tasks (i.e. those that involve one way of colouring-in, copying or completing) are 'seen as end results or "products": such products fail to engage children in connecting internal and external representations at a deep level. They promote "production" of "representations" that lack meaning and from which no relational statements can be drawn'. In contrast, 'in more realistic learning contexts, students may make sense of complex phenomena through their efforts to construct and through the use of graphical representations of these complex systems' (2000: 125).

However, the significance of children's personal journeys into the written language of mathematics has until very recently rarely been acknowledged in early childhood curricula (Carruthers and Worthington 2006). We believe this is likely to have been the origin of at least some of the disinterest, disaffection and dislike of mathematics for many children throughout their school careers – and for the high percentage of adults in England for whom mathematics is 'the elephant in the classroom' (Boaler 2009). The result has been a widely accepted 'can't do' attitude to mathematics in England (DCSF 2008a: 71).

Multimodality

Children's symbolic play and graphicacy can be viewed as different 'modes' or ways of representing meanings. Gunther Kress has investigated the relationship between children's *multimodal* meaning-making – with various media and artefacts, junk models, drawings and cut-outs (1997) – and the very foundations of literacy (2003).

Kress's seminal work on multimodality is also rooted in semiotics: the foundations of multimodality can be traced back to Vygotsky's research in the 1930s on symbolic play and children's meanings. The growing body of research in this field is where our book begins.

In early childhood, multimodality has been studied from various perspectives including combinations such as pictures, gestures and gaze (Lancaster 2001), and the syntax of graphical marks (Lancaster 2007), reminding us of the sand- and stick-talk described earlier. However, Flewitt emphasizes that while it is now recognized that children make meanings through everything they do, in the current educational climate only certain modes are valued. She argues that it is the more easily 'assessable' modes of spoken and written language which are prized and that 'the multi-modality of pre-school children's meaning making remains undervalued and under-researched' (2005: 209).

Why maths matters

It is widely agreed that mathematics and literacy are highly significant aspects of education. They enrich our understanding of the world, enabling us to develop and communicate our ideas and apply them in our social lives and, in the long term, they contribute to our success in work and in society at large. 'Success' should not only be measured in terms of achievement in examinations, but also in terms of individuals' interest, motivation, confidence and enjoyment – elements that are important throughout childhood and indeed throughout life. Yet for all its importance it is clear that many children continue to experience problems in their experiences of learning mathematics. Boaler (2009: 36) argues that the difficulty lies in the way in which mathematics is taught:

> Students are forced into a passive relationship with their knowledge – they are taught only to follow rules and not to engage in sense-making, reasoning, or thought, acts that are critical to an effective use of mathematics. This passive approach, that characterises maths teaching in many schools, is highly ineffective.

It is in their earliest educational experiences of mathematics that children develop their personal attitudes and beliefs about the subject and about themselves as young mathematicians. Dweck and Leggett (1993, cited in Sylva and Wiltshire 1993: 32–3) showed that teachers working with children in the early years contribute to their personal beliefs about their ability to succeed, and that when meeting problems or likely failure, children respond with one of two different patterns of behaviour: either 'helplessness' or 'mastery'. This also raises the vexed question of 'setting' for mathematics that in England is often used to group children by 'ability', sometimes from the age of 4, yet which 'has no positive effects on attainment but has detrimental effects on the social and personal outcomes for children' (Blatchford *et al.* 2008: 28). Moreover, Boaler (2009: 95–7) argues that:

> In England we do something that is very cruel to children in mathematics classrooms that sets us apart from just about every other country in the world. We tell children from a very young age, that they are no good at maths . . . deciding that primary age children have 'low ability' and grouping them in a low set is damaging . . . and does nothing to raise attainment.

Challenges and attitudes

One of Boaler's main criticisms is that children in school are often taught to memorize formulae and apply them without understanding, yet in the 'real world' people rarely use calculation methods they have been taught in school, using instead personal methods (see e.g. Nunes *et al.* 1993). 'It is so important', Boaler argues, 'that schools develop flexible thinkers who can draw from a variety of mathematical principles in solving problems. The only way to create flexible thinkers is to give children experience of working in these ways' (2009: 50).

'Written' mathematics has been shown to cause young children considerable difficulties (e.g. Ginsburg 1977; Hughes 1986; Dowker 2004). It appears to suffer from a restricted pedagogy more than any other subject, related in part to anxieties that some early years practitioners have about mathematics. And, since mathematics is generally viewed as a 'hard' subject and children in early childhood education settings are very young, a commonly held view is that the 'skills' of mathematical notation (i.e. written numeral formation and written calculations) need to be taught directly.

Ernest proposed that curricula reforms depend on teachers' beliefs and that 'Empirical evidence suggests that teachers may interpret problems and investigations in a narrow way' (1991: 289). Furthermore, school mathematics is often viewed as having one correct solution (Lerman 1989), frequently resulting in 'transmission' teaching. Therefore, as mathematics can appear to be a straitjacket, an exact science with only one answer, and play often seems messy, unruly and with no particular purpose, they are often seen as polar opposites: for most the question might be how these two apparent extremes can be reconciled. Of course mathematical play (sometimes referred to as 'practical maths') is of enormous value as children explore concepts that underpin number, shape, space and measure, pattern and other aspects of mathematics though sand, water, puzzles and other resources. But these experiences are not *directly* related to children's developing understanding of 'written' mathematics, since to learn this children need to 'write' mathematics in ways that make personal sense.

A study of teachers' attitudes towards mathematics showed that while they recognize the importance of mathematics for their daily lives, work and society, it was only those who felt open and confident about mathematics themselves who placed the greatest emphasis on the *processes* of learning the subject. In contrast, teachers who 'feel reluctant towards mathematics . . . undervalue the aspect of process'. They have a 'traditional' approach to their pedagogy as van Oers (2001: 27) observed at the beginning of his study '. . . those teachers have the children spend most of their time just working on worksheets . . . they have to colour shapes, count or encircle objects, etc' (Thiel 2010: 111).

The role and significance of graphicacy has been considerably undervalued in the early years, and until recently its importance has seldom been acknowledged in early childhood mathematics. This understanding has its roots in imagination and symbolic play, since this is where children begin to 'make meanings' (through gestures, actions, words, artefacts, models, marks or graphical signs), using them to 'mean' or signify something else.

In the current English Early Years Foundation Stage curriculum (EYFS) (DfES 2007a), the language area of the curriculum now known as 'communication, language and literacy' emphasizes its communicative role in the title, and has distinct sections entitled 'language for communication' and 'language for thinking', yet the section on mathematics (now entitled 'problem solving, reasoning and numeracy') fails to mention either. It is almost impossible to conceive of the 'language' area of any curriculum without these aspects. The same document emphasizes that building the foundations of literacy includes 'making sense of visual and verbal signs' (p. 39), yet this is also omitted in the mathematics section of the curriculum.

A new perspective

We have been researching *children's mathematical graphics* for the past 20 years. When we began in the nursery and schools in which we taught, our intention at that time was only to develop our pedagogy to better support the children's mathematical thinking and understanding. 'Children's mathematics' – mathematics that is real and personally meaningful and rooted in their play and social contexts – can overcome some of the problems cited here: and at its heart *children's mathematical graphics* offers a new perspective in supporting their understanding of the 'written' language of mathematics.

Our first book (Worthington and Carruthers 2003) includes examples from the children we had taught during a period of 12 years, beginning in the early 1990s. Since then we have worked with numerous children, teachers and practitioners in other settings and schools, through focused projects in schools and local authorities, through the medium of e-learning, on professional development courses and with students at colleges and universities. Teachers who have embraced *children's mathematical graphics* have generously provided all of the examples and case studies in this book, showing just what can be achieved. The observations and examples featured here have been gathered recently and demonstrate that teachers and practitioners in a wide range of settings, in different neighbourhoods, cultures and countries are supporting *children's mathematical graphics*. Learning is a social practice that depends on the interest, support and understanding of sensitive early childhood educators and the learning cultures they develop. The examples and case studies in this book reveal the professional journeys teachers and practitioners have made, providing insights into their considerable conceptual shifts and celebrating young children as powerful thinkers and meaning-makers.

Rather than focus only on children's *products* of mathematics, effective pedagogy (informed by research and theory, observation, reflection and understanding) can reveal the *processes* of children's mathematical thinking and the semiotic potential of play and graphicacy for maths. There is no justification whatsoever for children to be taught to a narrow set of criteria that are 'ticked off' against a predetermined list; nor to copy, colour or complete a worksheet or something 'mathematical' an adult has drawn or written. Such practices fail to excite children about mathematics and under-value their potential: they simply won't do. Nutbrown observes that 'children approach their learning with wide eyes and open minds, so their educators too need wide eyes and open minds to see clearly and to understand what they see . . . to see what is really happening and not what adults sometimes suppose' (2001: 134).

As the play episodes and numerous examples in this book show, there are other, richer interpretations of 'written' mathematics that arise through child-initiated play in naturalistic and democratic contexts of early childhood settings, and that show how rich with potential such play (and mathematics) can be.

Conclusion

The practitioners who have generously shared their observations and examples with us have such 'wide eyes and open minds'. We hope that readers will share with us our delight in and respect of the power of young children's profound capacity to make personal meanings as they make sense of their worlds. The themes explored in this introductory chapter – play, multimodality, meaning-making and graphicacy, underpin *children's mathematical graphics* and are woven throughout this book. In the next chapter we reveal something of this complexity as children explore their imagination and symbolic play.

Reflections

- Discuss your own feelings about mathematics with your colleagues: have those feelings influenced the ways in which you teach the subject?
- What mathematics do you see in children's self-initiated play? Do children choose to spontaneously 'write' mathematics (i.e. represent on paper or another surface) in their self-initiated play?
- With your colleagues, discuss some of the ways you have seen children represent their mathematical thinking (have you seen children do this)?

Recommended reading

Boaler, J. (2009) *The Elephant in the Classroom: Helping Children Learn to Love Maths*. London: Souvenir Press.

Simpson, J. (2006) Sand talk and how to record it, in *Transient Languages and Cultures*, http://blogs.usyd.edu.au/elac/2006/10/sand_talk_and_how_to_record_it.html, accessed 17 May 2009.

Broadhead, P., Howard, J. and Wood, E. (eds) (2010) *Play and Learning in Educational Settings*. London: Sage.

2
Beginnings in play

'Brothers are boys, girls are girls.' 'You're a girl, Margaret.' 'So are you, Mollie, L-M-N-O. That spells "girl".' 'You don't know how to make a triangle,' Mollie retorts . . . 'Yes I do! Teacher, Mollie says I can't make triangles!' . . . 'I do know how. And I can make a hexigon and you're not my friend Mollie.' 'Yes, you can make a triangle, Margaret,' Mollie says. 'You really can . . . Let's be kittens, okay?' 'Are you a hexigon kitty?' Mollie asks, setting the table.

(Paley 1986a: 35)

What is this chapter about?

- **Imagination and symbolic play**
- **Meaning-making** – symbolic tools; meaning-making and mathematics
- **Understanding the functions of imagination** though a number of case studies
- Showing how the **mathematical potentials of imagination** relate to *children's mathematical graphics*.

This chapter reveals something of the complexity of play as children explore and make meanings through their imagination.[1] We show how symbolic play can support abstract and divergent thinking and its potential for children's understanding of the abstract symbolic 'written' language of mathematics.

Imagination and symbolic play

Children engaged in symbolic play make (represent) meanings and this has been shown to offer potential for children's understanding of symbolic languages such as

---

[1] This chapter draws on a chapter written by Worthington, M. (2010a) Play is a complex landscape: imagination and symbolic meanings, in P. Broadhead, L. Wood and J. Howard (eds) *Play and Learning in Educational Settings*. London: Sage, used with kind permission of the publisher.

writing (Vygotsky 1978) and the written notation of mathematics (van Oers 2005; Worthington 2010e). The word 'imagination' means *image formation*, in which, 'by making and using signs, people make images of their reality' (van Oers 2005: 5). Pramling (2009: 273) argues that 'Human knowledge consists to a large extent of representations'. Children's images or signs include a diverse range of gestures, actions, sounds and words, artefacts and graphicacy.

Vygotsky viewed imagination and symbolic play as contributing to 'the highest level of pre-school development' (1978: 102–3). Emphasizing the importance of children's meanings, he argued that 'each step' supports the role of imagination in cognition, resulting in cognition becoming 'more complex and richer' (1987: 349); imagination 'does not develop all at once, but very slowly and gradually evolves from more elementary and simpler forms into more complex ones' (2003: 12). As 'children continuously weave in and out of play', they transfer 'real world' knowledge, skills and understandings from other areas of their lives. For example, in their symbolic play children are intent on making and exploring personal meanings about roles, situations and artefacts from their social and cultural experiences, something that Moll *et al.* refer to as 'funds of knowledge' (1992).

Kress argues that, 'We know that young children are clever, even wise, at an early age; yet we may shrink from attributing full intentionality to the things they do and make, so effortlessly, and which look so unlike our adult conceptions of what things are and should be' (1997: 35). The meanings and intentions of young children's representations are not always immediately transparent or accessible to adults – for example, their drawings and models 'are not usually subjected to the same analysis for meaning, not seen as much a part of *communication* as language is for instance. Images of most kinds are thought of as being about *expression*, not *information, communication*' (Kress 1997: 36, original emphasis). This suggests why children's 'art' and junk models are so often bracketed under 'creativity' or (later) by adults as an afterthought – something to be added to a piece of writing.

Meaning-making

Symbolic tools

Children use various media and resources to explore and represent their ideas through 'signs' or symbolic tools, and these semiotic resources are rich with potential meaning. Just as physical tools (such as hammers or spades) enable us to solve physical problems, meaning-making (semiotic) activities result in *symbolic* tools that can be used to resolve particular mental problems (Vygotsky 1978). We make meanings through 'signs' and 'texts': from a semiotic perspective these refer to much more than written (or print-based) materials and include all symbolic tools, whether, in the context of children's play, they are artefacts used to signify a particular meaning; or actions, spoken words, models or graphicacy. In respect of young children's play, Pahl describes children's 'texts', including arrangements of items, models and cut-outs, as 'visual, textual and artefactual practices' (2002: 145), as children 'use such representational means as they have available for that meaning making' (Kress 1997: 17).

The term 'multimodality' refers simply to the many ways in which meanings are explored, made and communicated with many materials, with what Kress (1997) refers to as 'lots of different stuff'. The choices children make within their play lead to the creation of these 'signs' and 'texts' so that arrangements of objects, the use of space and the many other decisions children make serve to influence and shape their signs – i.e. the 'design' of their signs (Kress and van Leeuwen 2001). Many signs and texts are complex 'hybrids' in which children combine various modes of representation, for example as in Mason's 'switches for Power Rangers' (see Figure 2.7) and can also occur within their graphical representations, as in Aman's 'boat-water' symbol (see Figure 4.3). In these, '*holistically*, in a single form . . . the meaning is constituted by its total effect and understood as a complete whole' (Wright 2010: 13–14).

The research that underpins this book is rooted in cultural-historical theory (also referred to as sociocultural theory). From a Vygotskian perspective, symbolic play clearly supports children as they create and use a range of symbolic 'tools' in their graphicacy: 'as in play', Vygotsky argues, 'so too in drawing' (1978: 110). As they assign meanings to their marks and symbols, children literally 'make meanings', and some of these will be mathematical. Kress (1997) views children's sign-making explorations as '*before writing*': not as 'pre-writing' activities but as rich play through which children explore and communicate complex meanings that support and reveal their early understandings of how writing 'works'. Kress shows how visual media increasingly influence children's understanding of print (writing). Kate Pahl (1999a) further explores a range of ways in which young children make meanings, providing rich insights into, for example, junk models and cut-outs. Other researchers have highlighted the relationship between popular culture, new media, new technologies and literacy (e.g. Holland 2003; Marsh 2005a), exploring how these technologies and 'super-heroes' influence and enrich children's literacy and play.

Pahl's richly detailed and analysed observations of young children's multimodal explorations enable us to see the tremendous diversity of their meaning-making through the models they make (1999a, 1999b), and in other research she has also focused on maps, ephemera and arrangements of artefacts and literacy practices (2002).

Exploring their 'texts' from this wider and more generous social-semiotic perspective allows us to see children's tremendous cognitive ability as they make diverse 'signs' and 'texts' within social contexts. In their self-initiated play, children choose the resources or media that best suit their purpose, and the culture of their setting can make a considerable difference to the extent to which children will choose to do this.

Meaning-making and mathematics

We acknowledge that children's meaning-making develops within their symbolic play in the first dimension of our taxonomy of *children's mathematical graphics* (see page 76). For example, in the following case study Sophie and her friends used a variety of artefacts to signify various items of food as they played 'picnics' (Worthington (2010a: 135)).

CASE STUDY

Sophie's picnic

In the nursery, Sophie was playing with several friends and hid various artefacts in a box of hay. Felicity held out a toy wheel, naming it as a 'chocolate cake' and others followed her lead, naming various cylindrical objects as 'chocolate' and 'strawberry' cakes. Sophie suggested they have a picnic and handing out wooden blocks as 'ice creams' gave a large seashell 'to eat' to an adult nearby. Lifting the shell to her mouth the adult paused asking if she should eat it, adding 'It's very rough and might hurt my mouth.' Sophie hesitated, watching warily as the adult put her finger inside the shell and licked it, then smiled and remarked that it tasted 'delicious'.

It was clear that Sophie was anxious to see if the adult understood the 'rules' implied in their spontaneous game: would she spoil it by saying she couldn't eat a seashell? Did she understand that for that moment it was important that all the players in the game shared the same 'rule' – that the objects they employed signified food? The players in the picnic needed to have a shared understanding of the rules and meanings in this play: the adult's action (pretending to scoop out some ice cream and eat it), her use of language, her tone of voice and her smile as she commented 'delicious' allowed meanings to be negotiated and confirmed that the adult understood and was able to share in the children's pretence and sustain play. Finally, as if to affirm acceptance of the adult as a full player in their game, Omar reached for the shell and said, 'Let's have some!'

Children's self-chosen and spontaneous imagined scenarios, and their enacted 'super-hero' play are rich in possibilities because they are the children's own (Worthington 2010c) and of their cultures (Riojas-Cortez 2001). As adults we might perhaps see only a child playing with a shell, but if we listen and carefully observe the child's play we may understand more of their world and uncover their personal meanings.

A direct relationship exists between children's meaning-making in play and their ability to use marks and symbols to mean (or signify) something, as the examples in this chapter show. This symbolic play is the precursor of symbolic languages such as writing and mathematics: in their *mathematical graphics* children may combine signs such as scribble marks, drawings, letters, words and numerals, along with personal (invented) and standard symbols to convey mathematical meanings. Combinations of marks, letters, numerals, symbols and drawings arranged in personally chosen ways on the page create texts that can be 'read' for their meanings.

Conceptualizing imagination

Van Oers argues that the 'products' of semiotic activities in young children's play gradually evolve '*in close relationship to the function it serves in different contexts*' (2005: 9, emphasis added). He investigated the 'potentials of imagination' and identified two categories or 'functions of imagination' based on empirical evidence, which underpin divergent and abstract thinking: these 'potentials' include the 'beginnings of mathematical imagination' (2005: 15–11).

Understanding the functions of imagination

Imagination as et cetera-act

Van Oers explains 'imagination as etcetera-act . . . actually refers to the invisible, by suggesting – with the help of some symbolic means – that a given series or rule can be continued'. He provides an example of a girl who explained her drawing by adding additional information beyond what she had drawn: the additional information was *implied*. These functions of imagination 'combines different elements' from the child's daily life 'into a new configuration' and reveals the child's 'alternative view of the world', helping us to understand 'the function of imagination and creativity': it suggests a rule or series that can be continued – a form of abstract thinking (2005: 8). Recent research in nursery settings has identified a number of examples of imagination as et cetera-act and some of these are included in this chapter.

Recalling group discussions about our bodies and skeletons, Finley later chose to make what he described as a 'puppet' (see Figure 2.1), explaining that the cuts across the fold of the paper plate represented 'ribs' and other lines were 'legs'.

Mima folded and snipped a piece of paper and, noticing the serrated edge created by the pinking shears she remarked with surprise 'A crocodile!' (see Figure 2.2). Kress (1997: 87) provides an example of a 4-year-old who noticed the slice of toast her father had bitten: 'You made it like a crocodile!' she said.

In both of these examples, the details are *implied*: rather than attempting to make a precise model, the children used economical signs to capture some of the essential elements in personal ways. In all of these signs that involve cutting, the outcome is a significant semiotic move from two to three dimensions that subsequently allows animation, making them 'real' (Pahl 1999a: 39).

Nathan used found materials to make an 'astronaut': taking a white envelope for the astronaut's suit he tucking coloured paper beneath the flap of the envelope to signify the astronaut (see Figure 2.3). Securing the flap of the envelope with masking

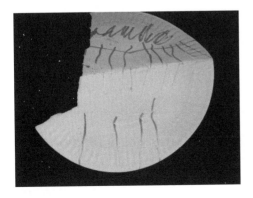

Figure 2.1 Finley's puppet

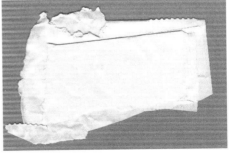

Figure 2.2 Mima's crocodile

Figure 2.3 Nathan's astronaut

tape allowed the astronaut to climb out of the space suit. Nathan accompanied his spoken explanation with actions, moving his model rapidly above his head in a trajectory to 'the moon', saying 'Blast off!' and making a *whooshing* sound as, in his imagination, the rocket left Earth. Gradually forms and meanings became discernible to adults' eyes.

Another example of *imagination as et cetera-act* comes from Hamzah, who drew a series of tiny circles (as wheels) to signify 'cars' (see Figure 2.4). It was only towards the end of his drawing that he added two black wheels beneath the row of grey dots on the left. Finally he drew the rectangular shape (lower right), with its two wheels and, pointing to the black dot inside the front of the car, explained 'driving wheel'. Hamzah explored his ideas about the particular features of cars that were important to him at the time. Kress includes a similar example from his 3-year-old son who drew a number of circles, focusing 'on that aspect of the object to be represented . . . representing these features of car, namely wheels or "wheelness"' (1997: 11). Kress

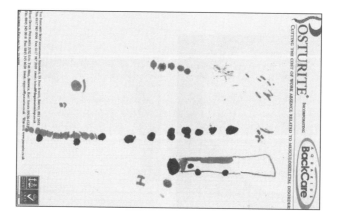

Figure 2.4 Hamzah's cars: see also Figure 3.3

emphasized that such signs are 'always transparent to their makers' (1993: 180), although without rich observations and collaborative dialogue with children it can be difficult for adults to understand some young children's intentions when they represent their ideas. It is important to recognize how observations are supported by knowledge of the children: their interests and cultures can illuminate our understanding. In all these examples the children had *implied* meanings: some added verbal explanations that supported adult understanding, and for very young children just one word may suffice.

Imagination as an act of generating alternatives

The second category van Oers identified is *imagination as an act of generating alternatives*, consisting of 'picturing how the world *could* be. These alternative representations are mental constructs' (2005: 9). He provides examples from Vygotsky (1978) of the child who used a stick in his play of riding a horse, and of another who used a wooden block to 'brush' a child's hair (Oliveira and Valsiner 1997). Many of the artefacts children appropriate or create in their play can be viewed from this perspective. For example, Mason and Mikey were playing outside and, finding some stones and logs, placed them around the perimeter of the large lorry tyre in which they stood, telling children and adults nearby about the 'treasure' they had found (see Figure 2.5). In a similar way, Sophie used toy wheels and some shells as *alternatives* for imagined food in her 'picnic' play (see page 16).

Three-year-old Tore was drawing excitedly on a whiteboard, explaining his marks as a 'shark' (see Figure 2.6). It seems possible that the marks suggested to him the rapid movement of a shark moving through (or vigorously thrashing in) the water; perhaps the dots and short vertical lines suggested 'teeth', although without further explanation of course we cannot know. Tore's drawing can be described as 'fortuitous realism' (Luquet 2001) in which the child notices something in the marks he or she has made. It suggests that 'Far from being chaotic actions and random

Figure 2.5 Finding treasure

Figure 2.6 Tore's shark

"scribblings" 'children's use and organisation of visual media exhibits semantic and structural characteristics from the beginning' (Matthews 1999: 90). Tore generated an *alternative* view of a shark and this allowed him to explore something he was excited about (and possibly a little afraid of): we can see that his drawing allowed him to explore his feelings in a safe context where *he* was in control of the shark. Other examples of this category include Nathan's 'writing (Figure 4.4).

Imagination as an act of dynamic change

In a number of examples the child appears to be the central player, controlling and exerting a direct influence on what happens next. Such observations differed from the two functions of imagination explored above and led to an additional function, *imagination as an act of dynamic change* (Worthington 2010a). The children assigned specific powers to their signs (either technological or occasionally magic), using them to 'operate' imagined technologies to effect change. Imagination as an act of dynamic change can be recognized as dynamic action, change, processes and transformations.

While this is a newly developed category it is certain that children have played out such themes and actions in the past. For example, Sybil Marshall recounts how, when reading part of a story written and illustrated by a boy called Jeff about an astronaut (also named 'Jeff')', she noticed that 'the real Jeff . . . was pressing with all his might an imaginary button on the side of my desk, after which he held the pit of his stomach and gazed at the school ceiling. "Gosh," he said, in an awed voice, "can't you just hear it?"' (Marshall 1968: 143).

The boundaries of what we understand to be 'play' constantly shift the more we observe, listen and reflect. Children's meaning-making embraces the many new technologies and media which are part of their culture (e.g. Kress 2003; Marsh 2005a; Marsh *et al.* 2005). The following examples of *acts of dynamic change* reveal that many of the children's invented technologies are based on real technologies they have operated in their homes, such as calculators, console and computer games, televisions and remote controls, and on the influences of new media and characters from popular culture.

Switches for Power Rangers[2]

Media appears to 'provide scripts for acts which then may or may not be materialized in everyday life' (Marsh 2005b: 45) and early childhood teachers and practitioners will often see children explore such themes in their play, as in the following play episode.

---

[2] The examples 'switches for Power Rangers' (Figure 2.7), and Mason's 'spy gadget' (Figure 2.9), along with 'paper calculators' (Figures 7.25 and 7.27) are included in Worthington, M. (2010) 'This is a *different* calculator – with computer games on': reflecting on children's symbolic play in the digital age, in J. Moyles (ed.) *Thinking About Play: Developing a Reflective Approach*, Maidenhead: Open University Press, and are used here with kind permission of the publishers.

Figure 2.7 Mason's switches for Power Rangers

Power Ranger play

Several boys were excitedly running indoors and out, engaged in playing 'Power Rangers' as they swung from a tree and climbed on and through a tunnel. Noticing some red wooden bricks, Mason tucked one in the end of his sleeve so that it was partly hidden, then he added several more bricks and inserted some in his other sleeve. Mason urged his teacher, 'Press that one and I turn red; that one – green and this one – blue! I'll leave them on until I see monsters!' Then, pressing one of the bricks, he shouted 'Kill!' After a short interval with two friends in the hammock, Mason climbed out and pressing one of the bricks announced, 'It's morning so I have to turn the lighting speed up!' Next he added another brick to his sleeve and admiring the effect said, 'This looks like a real Power Ranger one,' before running off.

Marsh *et al.* (2005) propose that children's 'media icons' are clearly very important: 'parents often described their children as being "obsessed" with them and were aware of the pervasiveness of their consumer culture' (2005: 45). Marsh and her colleagues found that boys in particular drew on their interest in new media and technologies such as televisions and remote controls, computers, game consoles and electronic toys, combined with popular culture and super-heroes.

Nathan's magic watch

During his early weeks in Reception, Nathan watched as Max cut a narrow strip of paper and, making marks on it, announced he'd made a 'watch'. Intrigued by this, Nathan took Max's idea and made his own, naming it a '*magic* watch' that enabled magic to be channelled through his fingers to the object or person to which he pointed and thus to effect some change (see Figure 2.8).

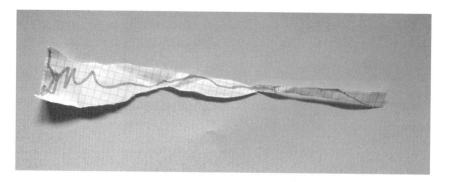

Figure 2.8 Nathan's magic watch

Mason's spy gadget

Children's multimodal sign-making (whatever the media) depends on their almost unconscious 'designs' chosen in terms of 'what is needed now, in this one situation, with this configuration of purposes, aims, audiences, *and with these resources*, and given *my* interest in this situation?' (Kress 2003: 49, original emphasis). Pahl argues that children's models play a significant role in their meaning-making and that this meaning-making 'needs to be carefully watched. It is often different from what we suppose' (1999a: 17). For example, early in his Reception class, Mason watched a girl nearby as she folded and snipped cuts around the perimeter of a piece of yellow card. Clearly intrigued by this Mason copied what she had done, but whereas Leola had made a card with a message to her mum, Mason's imagination took him into the world of contemporary film heroes and computer games, and he invested his gadget with special technological powers and drew on his considerable knowledge of passwords and controls (see Figures 2.9, 2.10). Taking a pen he wrote a string of letters and numerals on his card, reading 'sk' '714bp10' and, lifting it to his face, explained it was 'a spy gadget'. ' "Sk" is to keep the password safe. To switch it on you have to say "714bp10".' Asked if there was a way to switch his 'spy gadget' off, he replied excitedly, 'Yeah! You have to read it backwards!' promptly reading, '10 pb417'. Mason's spy gadget is a good example

Figure 2.9 Mason's spy gadget

Figure 2.10 'This is how it works'

of *intertextuality*, combining symbols (cut and folded card, a stamped clover-leaf and his string of letters and numbers) and 'weaving' these in an interrelated whole so that it contains his meanings and enabled him also to wear it (by covering his face). We can guess that when he held his spy gadget over his face Mason assumed a new identity. The letter-number string allowed him to 'assume a privileged status' in his class (Wright 2010: 54) due to the fact that handwriting, phonics and writing numerals have high *representational status* and are prioritized. But it is only through closely observing and reflecting on rich observations of children's play that we can see where their cultural worlds and interests meet: as Dyson's work (1997, 2001) has shown, children 'already engage in recontextualizing processes which link textual practices derived from a diversity of social worlds to the official school curriculum' (Millard 2006: 250).

Marsh *et al.* (2005) emphasize that these early 'digital beginnings' are heightened by the popular culture (e.g. characters, games) that populates all aspects of young children's lives and contributes to further shifts in their literacy practices. Marsh (2004: 63) argues that since children have access to an extensive range of contemporary 'texts' and technological artefacts such as television, computer and video games, virtual worlds and mobile phones, it is important that we acknowledge and understand these technological practices and their influence on children's literacies: 'In these contemporary communicative practices, we see the powerful "material objects, traces, and leavings" (Gell 1998: 222) . . . appropriate for the material, cultural and economic conditions of the 21st century' (Marsh 2004: 63).

Mathematical potentials of imagination

Created within their imaginative play, the 'products' of such semiotic activities gradually evolve 'in close relationship to the function [they serve] in different contexts (van Oers 2005: 9). Van Oers proposes that the et cetera form (abstraction) and the alternatives producing form (divergent thinking) are closely linked: in these he identifies the 'beginnings of mathematical imagination'. We find that the same functions (with the addition of *imagination as an act of dynamic change*) can also be traced in *children's mathematical graphics*. At the same time it is important to recognize that play should be valued for the many rich experiences and qualities it provides young children at the moment in which they are involved, rather than as a preparation for 'school' mathematics.

This analysis of imagination and symbolic play has uncovered some of the children's complex meanings as they made, adapted and co-constructed signs to fulfil specific semiotic needs. We have previously identified several specific aspects of the use of symbolic tools in mathematics: 'implied symbols', alternative ways of representing symbols (i.e. early written numerals), 'narrative action' and code-switching' (e.g. Carruthers and Worthington 2006, 2008): none of these signs had been taught. The following examples are from children of 4 and 5 years of age and show the relationship between the functions of the imagination and specific symbolic tool-use within *children's mathematical graphics*. They include alternative representations of numerals and symbols, invented and 'implied' symbols within calculations and, significantly, 'narrative action' (the use of hands and arrows to denote the active operation of calculation).

Imagination as an et cetera-act (abstraction)

The potential of this function is revealed in children's use of 'implied symbols' (symbols that are not represented but are present in the child's reading of their calculation) in their personal representations of calculations.

Related examples of children's mathematical graphics

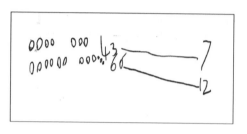

Figure 2.11 Jack adding grapes: *implied addition signs*

Figure 2.12 Jax adding grapes: addition and equals signs *implied*

Imagination as an act of generating alternatives (divergent thinking)

This is evident when children use alternative (personal) ways of representing numerals or mathematical symbols such as '+', and their own (alternative) ways of representing calculations or data.

Related examples of children's mathematical graphics

Figure 2.13 Alex's numbers: *generating alternatives*

Figure 2.14 Louisa adding strawberries: *generating alternative 'signs' for addition and equals*

Imagination as an act of dynamic change

This function appears to support children's understanding of how mathematical signs (operators) are used to effect change within calculations. It is evident when children use 'narrative action', representing arrows or hands to signify the action (operation) of subtraction as 'taking away'. Van Oers and Poland refer to children's use of arrows to denote change as '*dynamic*' representations: they too found arrows 'particularly important for the development of academic (e.g. mathematical) thinking' (2007: 15, emphasis in the original).

Indexical signs include *pointing to something* (with an *index* finger) and the use of *arrows*, which are distinctive visual signs such as those in road signs and technologies: they are 'a kind of linguistic pointing' (Mercer 2000: 23). Examples in this book include Nathan's magic watch (Figure 2.8) and James' explanation of his complex drawn narrative about a robot game (Figures 3.7 and 3.8).

Related examples of children's mathematical graphics

Researching semiotics in mathematics, Ernest describes three components of meaning-making systems: 'a set of signs . . . a set of relationships between these signs' and 'a set of rules of sign production for both single and compound signs' (2006: 409–10) and in these examples of *children's mathematical graphics* we can identify all three components.

Analysis of data in the current study reveals children's communicative competences and provides insights into the relationship between symbolic play and mathematical notation. *Imagination as an act of dynamic change* suggests a potentially valuable relationship with children's ability to understand their active role in operating on two quantities in calculations. The findings suggest that the semiotic 'potentials' of imagination support the emergence of *children's mathematical graphics* in ways that we are only beginning to understand.

Figure 2.15 Barney, subtracting beans: *narrative action* (also alternative signs for subtraction and equals)

Figure 2.16 Fred's addition: *narrative action* (also implied addition and equals signs)

Conclusion

Moyles argues that 'Although the word "play" is used in early years documents and policies in England, it is neither clearly defined nor well understood by practitioners or policy-makers in this country or in many other countries' (2010: xi). The provision and pedagogy of play appears to be a particular issue in many Reception classes, where research has argued for greater opportunities for 'sustained, complex imaginative play' (Adams *et al.* 2004). Recent empirical evidence suggests that this situation continues in 2010. However, since the current guidance for the EYFS devotes less than a page to the subject of play this may not be surprising (DfES 2007a). Wood argues that 'achieving good quality play in practice remains a considerable challenge, particularly in the UK where teachers face competing notions of what constitutes effective teaching and learning' (2009: 29). Wood and Attfield (1996) emphasize that practitioners need a more conscious, clearly defined basis to their pedagogical knowledge that makes explicit the links between play, teaching and learning. However, this may well be 'a contentious area, because of the ideological commitment to free play and free choice (Wood 2008, cited in Wood 2009: 27).

In her book on the role and status of play in early childhood education, Moyles questioned whether children could ever be said to be '*just* playing' (1989, italics added). Wood and Attfield argue that 'Play can be regarded as deeply serious and purposeful, or trivial and purposeless. It can be characterised by high levels of motivation, creativity and learning, or relegated to little more than aimless messing about' (2005: 2). They argue that 'the ambiguities surrounding the definition of play have done little to substantiate *learning through play* or that a *play-based curriculum* is the best approach to supporting early learning' (2005: 2, emphasis added). This confusion can clearly cause difficulties for teachers and practitioners in early childhood settings, and nowhere more so than in recognizing and supporting children's mathematics in their self-initiated play.

In the next chapter we focus on graphicacy through children's drawing, writing and maps.

Reflections

- Make one or two observations of children exploring their imagination and symbolic play: your observations will need to include as much detail as possible about the context of the play, and what the children said and did.
- Consider how the children made and represented their personal *meanings* within this play. What do your observations tell you about the children's thinking?
- In the coming week you may like to focus your observations on specific aspects of children's imagination and symbolic play (e.g. role play and super-hero play, junk models, block play, drawing, writing or children's own maps). Meet together with your colleagues to discuss your observations and focus on the different ways in which the children made meanings.

Recommended reading

Kress, G. (1997) *Before Writing: Rethinking the Paths to Literacy*. London: Routledge.

Marsh, J. (ed.) (2005) *Popular Culture, New Media and Digital Literacy in Early Childhood*. London: RoutledgeFalmer.

Pahl, K. (1999) *Transformations: Meaning Making in the Nursery*. Stoke-on-Trent: Trentham Books.

3

Drawing, writing and maps

Barney prints the letter D on a small piece of paper and creates a story in order to use it. 'Spiderman steps on a D. Then R2D2 steps on a D. Then C3PO steps on a D ..' 'If you step on a D,' Eric suggests, 'you fall in a trap.' 'And you get out with a ladder,' Libby adds.

(Paley 1986a: 63)

What is this chapter about?

- **Graphicacy** and cultural influences.
- **Young children's drawing**
- **Writing**, including writing in more than one alphabet or script.
- **Children's map-making.**

In this chapter we explore children's meanings through their drawings and writing, highlighting some particular features and focusing on aspects that have significance for children's mathematical graphics.

Graphicacy

Graphicacy has been described as:

> the ability to understand and present information in the form of sketches, photographs, diagrams, maps, plans, charts, graphics and other non-textual two-dimensional formats. The information can be directly representative of what we see (as in photographs or in drawings) or more abstract – for example information which is spatial (as in maps, plans and diagrams) or numerical (as in tables and graphs).
>
> (Aldrich and Sheppard 2000: 8)

Graphics were among the earliest communication modes to evolve, 'predating numbers and writing by thousands of years and providing strong support for the idea that graphic images are a fundamental aspect of human communication' (Porascy *et al.* 1999: 104–5).

Wilmot (1999: 91–2) suggests that 'Children of today inhabit a multi-dimensional world and in order to communicate effectively in it they need the ability to utilise four forms of communication, namely, oracy, literacy, numeracy and graphicacy. Communication in graphic form requires an ability to both encode and decode spatial information and symbols'. While young children's learning is widely understood as holistic, recent research continues to address drawing and writing separately, and research into graphicacy in early childhood education focuses almost exclusively on drawing (e.g. Anning and Ring 2004). The exception to this is research into multi-modality (discussed in Chapter 2), which embraces children's diverse ways of meaning-making including graphicacy (e.g. Kress 1997). Significantly, there has been little research that explores the relationship between drawing and writing, and until now none that acknowledges the relationship between the various aspects of graphicacy, including drawing, writing, maps and *children's mathematical graphics*.

Vygotsky and Kress have made a strong argument for the relationship between children's symbolic play and literacy and it seems logical to us that there also exists a relationship with mathematical representations. Such representations are generally referred to as 'written' maths in spite of the fact that they often include drawings, geometrical shapes, charts and graphs; alternatively the term 'notation' is sometimes used (e.g. Brizuela 2004).

Recent research into young children's drawings takes a developmental perspective (Matthews 1999), whereas our focus is on ways in which children use their graphical marks, symbols and other representations to explore and communicate their meanings and narratives. This recognizes 'the importance of context' and the impact of children's cultural contexts which, combined with their social interactions with peers and significant adults 'also impacts on the drawing process and the meanings constructed and conveyed' (Light 1985, quoted in Einarsdottir *et al.* 2009: 218).

Cultural influences on young children's graphicacy

Drawing creates powerful symbolic 'tools'. These originate in human culture, influencing and shaping the choices and decisions that children make (Collins 1976; Vygotsky 1978; Potts 1996; Golumb 2002). Writing also has significant cultural and historical roots (e.g. Olson 1994; Robinson 1995). Historically, humans have developed different cultural practices and although culture is seen as 'a very recent evolutionary product', it is likely that we have a 'biologically-inherited social-cognitive ability to create and use social conventions and symbols' (Tomasello 1999: 216). Culture clearly influences the way in which we 'see' something in marks or images – for example, the constellation of the zodiac that we refer to as a lion or 'Taurus' is seen by some native South American peoples as a lobster (Gombrich 1960: 91).

Children's drawings were previously thought to follow a universal pathway of development as they moved from their earliest scribbles eventually to achieve realistic images (e.g. Luquet 1991; Sully 2000). However, the drawings these researchers investigated (from their work in 1895 and 1927) now look very dated and it is easy to see that the children were influenced by the culture and drawings of their time. Studies by a number of researchers have shown how culture impacts differently in diverse global cultures in terms of choice of subject, colour, layout, symbols and style (e.g. Alland 1983; Stokrocki 1994; Matthews 1999; Cox *et al.* 2001; Wilson 2002). Differences appear to be particularly marked in children's drawings of human figures, as examples from Namibia and Zimbabwe, and from some native Australians reveal (Cox 2005). Yet children's drawing 'is not defined in terms of a body of knowledge, planned . . . and simply transmitted to the learner. Nor is it tied to the transmission of any particular culture' (Matthews 1999: 163).

Rosemary Hill showed that when native Australian children from the Walpiri tribe attended non-native schools, their drawings showed symbols from their own culture intermixed with western influences (Hill 1996, cited in Cox 2005): in second-language learning this is known as 'code-switching' (see e.g. Cook 2001) and is a feature we have identified in *children's mathematical graphics* (Carruthers and Worthington 2006). Significant historical events such as wars and the 2004 Asian tsunami also impact on children's lives and become the subjects for their representations with sand, drawings and storytelling (Rousseau *et al.* 2005; McMahon *et al.* 2006). While drawing and writing are culturally situated practices, they nevertheless do not merely involve 'copying' what someone else has done. Rogoff (2003: 236) argues that:

> Although thinking is often regarded as a private, solo activity, cultural research has brought to light many ways that thinking involves interpersonal and community processes in addition to individual processes. The study of cognitive development now attends to more than the unfolding of children's understanding through childhood and attends to how people come to understand their world through active participation in shared activities as they engage in sociocultural activities.

Kress (2003) argues that in the past drawing and written print-based materials such as books, papers and advertising in magazines considerably influenced children's writing. In education the impact of this often resulted in requirements for children to produce a piece of writing to a certain format, with occasional drawing allowed after the writing was complete. Convention dictated that the spatial organization of writing was (in written English) top to bottom and in a left to right orientation, and mathematical notation largely followed this convention. However, the influence of screen-based media has challenged this, resulting in different ways of visually representing and communicating meanings (Pahl and Rowsell 2005, 2006).

Young children's drawing

Young children's representations rarely fit neatly into specific categories. Lowenfeld and Brittain argue that drawing combines 'diverse elements of their experience . . . In the process of selecting, interpreting and reforming these elements, children have given us more than a picture or sculpture; they have given us a part of themselves; how they think, feel and see' (1987: 2). For example, Max cut two small rectangles of paper and joined them at the top so that they could be opened and shut (see Figure 3.1). Rapidly drawing a face on the top piece, he announced his paper artefact

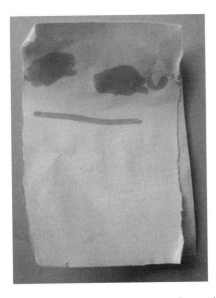

Figure 3.1 Max's card-girl (4 years, 7 months)

was a 'card-girl'. Although we might label this either a 'drawing' or a 'card', holding Max's tiny 'card-girl' in the palm of the hand also shows it is a hybrid artefact and suggests it could be moved and played with as a prop in his imaginative play – a small puppet or person.

Lancaster (2007: 129) argues that:

> Drawing tends to be seen as a transparent system: one that is easier to do than writing, and where you can literally see what you mean. The term 'drawing' is ubiquitous in adult interactions with young children where the making of marks is involved, often taking on an inclusive meaning such that it refers to any kind of mark that a child makes. The use of the term in this way suggests that drawing is viewed as some kind of innate skill; something that anyone can do.

Furthermore, whereas to adults young children's drawings may appear simple, 'drawing is as much a complex system involving rules and regularities as writing or mathematics' (Lancaster 2007: 130); in contrast, 'writing, recording numbers, or notating music tend to be deemed hard things to do, not innate, and therefore to require teaching' (Lancaster 2007: 129). However, the point we make is that where children are encouraged to use their own graphics, the meanings are theirs and will be deeply understood.

Scribbles

Vygotsky (1978: 113) argued that 'the child must discover that the lines he makes can signify something'. These beginnings are exemplified by scribble marks in Tore's 'shark' (see Figure 2.6) and by Finn's drawing which he referred to as 'night time' (see Figure 3.2): they suggest what Matthews (1998) refers to as 'action representation', while many of the other examples in this chapter suggest drawings of objects, or 'configurative representations'. Children's discovery of the potential of marks to signify meanings underpins all graphicacy including symbolic languages and is at the root of their *mathematical graphics*.

It is important to emphasize that in none of the drawings did an adult ask the child what they had drawn: the meanings they attached to their marks were offered spontaneously to either adults or their peers while they were engaged in their drawing, and were not based on any expectation that their marks should be 'about' something specific.

Hamzah loves cars and his drawing is connected to his feelings about journeys in the family car with his mother and father (see Figure 3.3). After representing his idea of the exterior of cars by drawing a series of circles to signify 'wheels' (see Figure 2.4), Hamzah turned to thinking about the car's interior. Pointing to the four sections within the grid, he explained that these were where members of his family sat: his father in the front (the lower-right section) with the 'driving wheel' and he and his mother in the seats at the back. Hamzah's drawing connects his personal identity within his family with positive feelings about going out together in the car. Like Tore, the drawing's subject can be seen as specifically gendered. Finally, Hamzah added two dots at the foot of his car for wheels, and wrote 'H' several times, the first letter of his name and an important feature of all young children's self-identity. Figures 3.4 and 3.5 are from two nursery children focusing on particular features of 'ships' and spiders – significant to them as they encode meanings, although their meanings may not always be readily accessible to us as adults. Both drawings are examples of what is known as *indexical terms* (to indicate something) in which, for example, a child enacts an explosion or uses signs to indicate this. An indexical term is also a way of communicating noise or movement (Wright 2010: 31), such as Nathan enacting the movement of his astronaut flying (see Figure 2.3), or Marshall's description of Jeff's spaceship (page 20).

Figure 3.2 Finn's 'night-time' (4 years, 1 month)

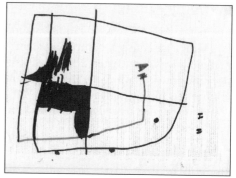

Figure 3.3 Inside Hamzah's car (3 years, 8 months): see also Figure 2.4

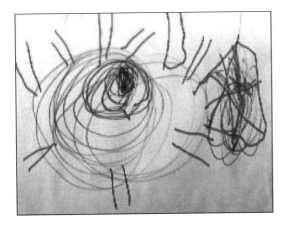

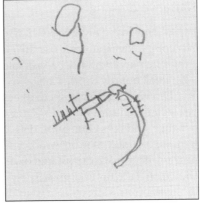

Figure 3.4 Max's 'ship that shoots' (3 years, 4 month)

Figure 3.5 Tore's spider (3 years, 5 months)

We have seen that young children assign meanings from their personal and cultural perspectives: this may be either through first-hand experience, for example of night-time, cars or spiders – or through secondary cultural experiences such as picture storybooks or television (e.g. sharks or a (space?) ship that shoots). In fact it is impossible to see that they could do otherwise. Gradually children begin to set out to draw with an idea in mind rather than attributing meanings *after* they have made their marks.

For adults, the content and meanings of children's drawings become more readily accessible as they mature, although features included often blur the boundaries between reality and imagination, as in Finn's drawing of a bull (see Figure 3.6).

Figure 3.6 Finn's 'bull with lots of horns and five legs' (3 years, 11 months)

Narrative drawing

James' drawings (Figures 3.7, 3.8) show how popular culture influences imagination. His complex narratives arose through his interest in 'robot (fighting) games' – console games he plays with his 10-year-old brother. James explained that he plays 'Level one – it's hard but my brother helps me.'

Drawing on a large sheet of card, James used a combination of intersecting lines, circles and figurative drawings to narrate the first 'episode' of his story. This is drawing *as* 'story' and is as equally valid as a written narrative. In the first drawing (Figure 3.7) James explained that his brother had drawn some of the figures (top left): the four nearest the top are ducks, one of which James drew. He explained, 'The ducks built a snowman' (the figure immediately below the ducks). The final figure drawn by James to the left of this group is 'a man drinking a milkshake – he's scared of the ducks and the snowmen!' The large circular shape in the centre is 'a house' and to the right of it the cross with extending lines is 'an aeroplane with things that go round' (propellers). The remaining shapes (most of which are identical to the aeroplane) are 'grenades to fight the king who lives in the house – to fight everyone!'

In his second, equally complex drawing (on the reverse of the first) James continued his narrative:

> The man drinking the milkshake is hiding – he's underneath and you can only see his eye [in the centre]. He's hiding from the aliens, he doesn't want to die. The electrics are blowing up the aeroplane and it crashed – it's wrecked. Some of the electrics are broken; some of the electrics are knotted up. The plane dropped grenades in the shark's mouth and on the houses.

It is interesting to note that both drawings cover the entire space, suggesting that James 'mapped' various aspects of his narrative onto different spatial locations of the paper. These examples contrast starkly with Finn's bull (see Figure 3.6) where the sky and the ground are placed horizontally at the top and bottom of the paper. James' drawings are highly complex visual narratives in which 'Each character is enacted

Figure 3.7 James' 'The ducks built a snowman . . .' (4 years, 3 months)

Figure 3.8 'The man drinking the milkshake is hiding . . .' (4 years, 3 months)

(spoken for) and narrated (spoken about)' (Wright 2010: 40) as he moved between characters and different events. Wright describes such narratives as 'loosely structured' and containing a range of evolving ideas: 'The themes that are depicted and enacted are like fleeting moments'. She likens *visual narratives* to films which move 'in and out of an overall plot scheme but [do] not conform to the conventions of having a clear-cut beginning, middle and end. Instead it is similar to fantasy-based play on paper' (2010: 45). The continuation of the narrative through the two drawings suggests distinct episodes of a story such as might be viewed on television, the plot unfolding over time.

Other examples of visual narratives include Megan's 'very big fast roller-coaster' (see Figure 3.9) and Max's story (see Figure 3.12) that includes both spatial and temporal qualities and suggests a map. However, it is possible that there are elements of visual narrative (in the child's mind) in the examples from very young children such as Tore's shark (see Figure 2.6) and Finn's 'night-time' (see Figure 3.2). James appeared to have been 'watching' his robot narrative unfold (rather than being within the story) but when the child positions herself within the picture as Nadieh did in her drawing of herself at the beach (see Figure 4.9), or Sophie and her friends enacting 'being at a picnic' (see page 16), they are playing out a *personalized narrative*.

We have already referred to the increasing evidence that modern technologies, new media and popular culture impact on children's play, drawings and model-making (Paley 1984; Dyson 1997; Pahl 1999a; Marsh 2005a), and James' drawings are examples of this. They show a significant masculine subject matter (e.g. console games, fighting and killing, planes and electricity) and suggest that he used these drawings as a means of addressing and controlling the rather frightening content of the computer game, thereby allowing him to gain some personal power over it.

Children's drawing at home is often free of adult expectations and agendas (Anning and Ring 2004). Figure 3.9 shows the first of three drawings by Megan about fairground rides, including a Ferris wheel and 'a runaway train'. Megan's drawing suggests the undulating route the roller-coaster took and its rapid movement, and includes its many seats. Megan told her mum, 'This is a very big, fast

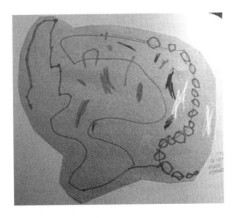

Figure 3.9 Megan's 'very big, fast roller-coaster' (4 years, 5 months)

roller-coaster!' Her mother explained that 'Megan was thinking about how much she'd love to go to a funfair again.' Megan's drawings suggest a *temporal nature* (in recalling past experiences) and she also recognizes the power of symbolic tools to communicate and persuade.

Writing

Children make choices and decisions about the graphical system they will use (e.g. drawing, writing, maps or *mathematical graphics*). Lancaster (2007: 130) argues that:

> When we talk about whether writing evolves from drawing, or is an essential precursor to writing, or simply ask whether children's early marks can be identified as drawing, writing, or enumerating . . . We are operating from sets of assumptions about graphic signs and systems, from our literate adult consciousness, that very young children cannot possibly share.

However, our research shows that at an early age children make and distinguish between the contexts and purposes of their graphical marks and visual representations. For example, in Figure 3.10 Liana explained she'd made 'a picture' for her mum. It began as a story might: 'It's a rainy day' to set the scene, continuing as she listed items. 'Here's a small flower, here's a bigger flower. This is a scarecrow; a house; a balloon and an apple.' Figure 3.11 also represents a list or inventory of items but in contrast Nathan used abstract 'written' symbols of crosses which he read as 'carrots, potatoes and spaghetti'. Nathan used conventional signs in a personal way, repeating the crosses as young children often repeat a known letter in their writing (Clay 1975). His

Figure 3.10 Liana's 'picture for my mum' (3 years, 6 months)

Figure 3.11 Nathan's 'shopping list': carrots, potatoes and spaghetti (3 years, 7 months)

mother explained that they always wrote a shopping list before the family went to the supermarket: Nathan had drawn on his home knowledge in his play at nursery.

The following examples all share features of inventories and all the children made personal decisions as they narrated their 'stories': Max (Figure 3.12) itemized 'A man holding a flag and a dinosaur with a hammer in his back-pack; and a mummy dinosaur with a bigger back-pack'. In contrast, pointing to the zig-zag lines she'd drawn, Jemima (Figure 3.13) explained, 'That's my writing and it says "b" for Mima' [Jemima's nickname] – and that's a little kitten with a belly-button.'

Elise was waiting for her mum who was serving customers in the local video shop. Her mum had given her a strip of the till receipt. She took the paper and wrote her name in the spaces, recounting a story about dinosaurs as she did so (Figure 3.14). After the dinosaurs in her story had a few 'fights', Elise 'read' 'The dinosaurs went home to have their tea, the end!' Elise had made her own connections between the print of the till receipt and stories. She also drew on her personal experiences of stories that are resolved with a happy ending as she 'wrote' and recounted her story.

Figure 3.12 Max: 3 years 3 months: 'A man holding a flag and a dinosaur with a hammer in his back-pack . . .'

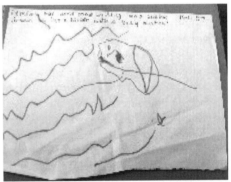

Figure 3.13 Jemima's story (3 years, 3 months)

Figure 3.14 Elise's dinosaur story (3 years, 10 months)

Writing in more than one alphabet or script

There has been extensive research stretching back over the past 40 years into young children's emergent writing, which began earlier with Vygotsky and Luria investigating the beginnings of early writing in the 1930s. Research in recent years has also investigated 'bi-literacy', showing how children of primary-school age develop personal theories about writing and build on what they know, drawing on their cultural understandings about writing in their first language to write in their second language: this includes the orientation, letters or script, the syntax and grammar (see e.g. Gregory and Williams 2000; Kenner 2004; Drury 2007). Kelly has studied young children's bi-literacy and she emphasizes that the children in her research 'were creatively blending understandings from their involvement with the different worlds of home and school' (2010: 86), and although this may appear to be particularly evident in children writing in two languages, from a cultural-historical perspective this blending of home and school cultural practices is evident in all children's symbolic representations.

While children of 3 and 4 years of age for whom English is a second language are unlikely to have developed mature writing in either their first or second languages, they already have awareness of the different appearances of the letters or characters of the scripts and the directionality of writing. The examples from Lay Hau Yun (Holly) and Auden (Figures 3.15, 3.16) are interesting in that the children appear to be establishing themselves as dual language writers, almost saying, 'Look! Let me show you that the two languages I speak are written in different ways!'

Holly's Vietnamese name is Lay Hau Yun and her home language is Cantonese. The example in Figure 3.15 shows how she represented her name twice, first as a zig-zag line (top left) as many young children do in English, and then on the right in Cantonese characters. On another occasion while playing, Holly was writing numbers as she counted, beginning '1' and '2' and following this with the written Cantonese character $\equiv$ for '3', integrating her knowledge of both written languages.

Auden's first language is English and his mother comes form Sri Lanka. He showed his teacher Mrs Sutcliffe her name in Sri Lankan and her 'proper name' (referring to written English) (Figure 3.16).

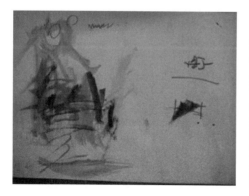

Figure 3.15 Holly (Lay Hau Yun) (3 years, 8 months)

Figure 3.16 Auden (4 years, 2 months) writing in two languages

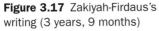

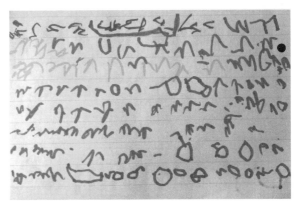

Figure 3.17 Zakiyah-Firdaus's writing (3 years, 9 months)

Figure 3.18 James: 'Seeing who's here' (4 years, 9 months)

Figure 3.17 shows Zakiyah-Firdaus's spontaneous writing in 'emergent' Arabic. Her brother attends classes at the local Koranic school and has written Arabic homework. To her mother's surprise Zakiyah-Firdaus had absorbed a great deal about written Arabic from watching her brother, including its right-to-left orientation. Her mother confirmed that the first two lines are from the Arabic alphabet and that further down Zakiyah-Firdaus had written 'Allah' in Arabic. On another day she wrote a name badge for her teacher, Emma, writing from right to left as she does in Arabic – but in emergent English, using letter symbols that her counting 'one, two, three' out loud and then continuing in Urdu: '*arba'a, khamsa, sitta, saba'a*' (four, five, six, seven). Emma noted that as she counted the numbers in Urdu, Zakiyah-Firdaus made gestures with her forefinger in the air as though she was writing each Urdu numeral, and checking how they are written seemed to confirm this.

In contrast, James (Figure 3.18) used a range of letter-like signs in his emergent writing (of English), written from left to right. He understands that writing in our alphabetic script uses a range of different letters and thought about some of their features. He also included short zig-zags as many young children do, perhaps encoding his sense of the appearance of cursive 'writing', or capturing the movement of an adult's hand as they write. James is a 4-year old and in his first term at school and 'Seeing who's here' was likely inspired by the register. Registers have personal meaning to young children as they think about their peers and identify each, enacting the teacher's role and at nursery and school writing 'registers' is often a preoccupation in children's play.

In Figure 3.19 Saja drew lines as she counted in Arabic: '*Wahid, ithani, thelaitha, araba, khumsa*' and then continued saying, 'five' and 'seven' in English. Signs in the learning environment in the children's other languages will support those whose first language is no English (Figure 3.20).

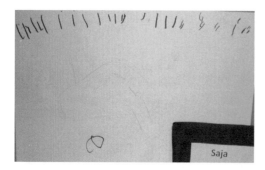

Figure 3.19 Saja's numbers (3 years, 5 months)

Figure 3.20 Multilingual signs in the environment

Maps

From a semiotic perspective maps offer insights into an intriguing genre of young children's graphical representations since they combine drawing, writing and other symbols to communicate very specific information.

Maps have always reflected the views of the mapmaker and the culture and society of the times in which they are made (see e.g. Chatwin 1987; Brody 1988; Brook 2009). For example, the oldest known map in the world, the Babylonian Mappa Mundi reflected the view of the ninth century BC that Babylon was the centre of the universe. Shown from a 'bird's-eye' view, it combines both real and mythological features, reflecting the Babylonians' knowledge of the world at that time and their cultural myths (Finkel and Seymour 2009).

Maps children choose to make for personal reasons have personal meanings, enabling them to explore their ideas about location and direction, space, position and relationships, distance, size and layout. These intertextual or 'hybrid' texts are a feature of young children's graphicacy and, as we show in the examples in this book, they often combine various marks, drawings, letters, words and other symbols to signify and communicate their meanings.

During quiet time, large sheets of paper were laid out next to a printed road map in the art area. Nathan made a large spiral pattern and, as he got nearer to the centre of his spiral, the lines drew closer together (Figure 3.21). He also added some small marks within the spiral. 'It's a road,' he explained, tracing his finger along the lines he had drawn. 'Here's the trees,' he added, pointing to the smaller marks at the centre of the spiral.

At small group time in the graphics area, Cameron made marks on a piece of squared paper (Figure 3.22). Holding a ruler on the paper he drew a line and then made several smaller marks. Turning the paper over he drew several large, circular lines that almost filled the page, and then using a compass with a pencil attached, made smaller marks within the circular shapes. 'That's the minibus,' he explained, pointing to a small shape he'd drawn. Then, tracing his finger along one of the longer lines, he showed where the minibus went.

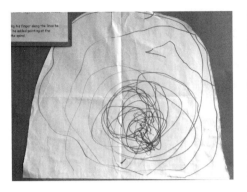

Figure 3.21 Nathan's map (3 years, 7 months)

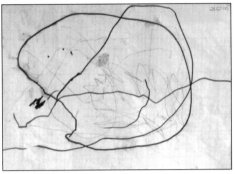

Figure 3.22 Cameron (3 years, 4 months)

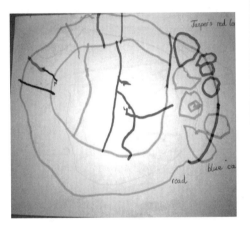

Figure 3.23 Jazper: 'Red lorry, blue car and a road' (3 years, 6 months)

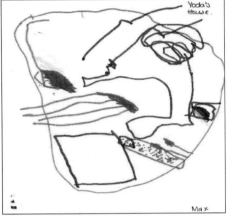

Figure 3.24 Max: 'Yoda's house' (5 years, 1 month)

Like James' 'robot game' narrative, images and ideas often appear to combine in ways that may seem unlikely to adults. For example, explaining his drawing Jazper (Figure 3.23) said, 'Red lorry, blue car and a road,' adding 'lots of eyes': his drawing seemed to combine elements of both drawings and maps. Map-making is always subjective and children centre their maps on their personal perspective such as a journey they have made, or link personal interests, as Max did in his map of 'Yoda's house' that relates to his interest in popular culture and the film *Star Wars* (Figure 3.24).

Pahl (2001) compared the maps children in her study made at home and at school. She makes the point that it was the home maps that particularly enabled the children to draw on 'an immense amount of cultural capital' (Bourdieu 1977), but that the maps' 'meaning-potential' could be realized and re-contextualized in school. She describes the children's maps in her study as 'inter-textual objects . . . played out over sites and across rooms, being at once a story, a narration, a type of play, a board game, a computer game and a map for gold' (2001: 124). Children's understanding of the graphical language of maps can be mediated through collaborative discussion with adults, and printed maps and globes provide additional symbolic tools that children may choose to incorporate in their own maps.

Conclusion

Since young children's personal views of the world are holistic, they do not 'see' learning in discrete 'subject' areas such as 'art', 'literacy' or 'mathematics' – as drawing, writing or 'written' mathematics. As they encode personal meanings children select and adapt from these different symbol systems, adapting and inventing symbols, fonts, syntax and layouts that best suit their purpose in complex and meaningful ways. Children use visual signs and texts in incredibly powerful ways and already by 3 or 4 years of age are able to select particular symbols for their contexts and audiences.

In the next chapter we focus on children's graphical signs and their contexts and meanings.

Reflections

- Observe several children as they draw or write and note down anything they do or say as they draw. Talk to the children about any aspects you'd like to understand more about. It would be helpful if your colleagues also did this.
- Meet together to discuss the children's graphics, focusing on the ways in which the children used particular marks, symbols and representations to communicate meanings.
- Observe children with English as a second language when they choose to write, noting in which language(s), alphabets and direction they write, and anything they say while they write.
- From a formative assessment perspective, discuss all the positive aspects of their writing. Consider the potential of discussions about children's graphics for formative assessment.

Recommended reading

Kelly, C. (2010) *Hidden Worlds: Young Children Learning Literacy in Multicultural Contexts.* Stoke-on-Trent: Trentham Books.

Matthews, J. (2003) *Drawing and Painting: Children and Visual Representation.* London: Paul Chapman.

Wright, S. (2010) *Understanding Creativity in Early Childhood.* London: Sage.

4

Graphical symbols: contexts and meanings

'Don't go there! Vijay. It's a poison river!' 'Did my feet get poisoned?' Vijay asks. 'Don't worry. I'm making you invisible. Touch this paper. Now you're un-poisoned. It's invisible writing.' 'What does it say?' 'That you're invisible. No one can see you except me.'

(Paley 2004: 60)

What is this chapter about?

- **Graphical symbols** – exploring contexts and meanings in children's drawings, writing and their mathematical graphics.
- **Flexible uses of graphical symbols** – including children's use of graphical symbols for mathematics.
- **Discussion and definitions of 'mark-making'.**

This chapter explores children's use of graphical symbols in their drawings and writing and provides a bridge to children's mathematical graphics, showing that children share meanings for certain symbols, and how flexible they are in their use of signs in a range of contexts.

Graphical symbols: contexts and meanings

John Matthews shows how the 'structures' of children's earliest visual representations are closely related to gestures, and are universal (1999: 33). These 'structures' have a timeless quality, appearing, for example, as rivers, snakes and lightning in petroglyphs and rock painting (e.g. Golumb 2002).

Matthews identified four 'generational structures' within which he grouped specific 'action-marks': he emphasizes that 'not only do young children use visual media to represent the structure and shape of objects; they also use these to represent the structure and shape of events' (1999: 31) (see Chapter 5). In addition to these generational structures, children are influenced by graphical symbols that have a strong visual impact and distinct uses in their culture, such as crosses and arrows. They make choices and decisions about the symbols they use to encode and communicate specific meanings and some of these become coded signs that others understand and accept. Lancaster explains: '[a] representational principle that emerges is that mark-types do not necessarily have fixed referents, but can be repeated and used in different environments, with the environment [context] being the significant factor in determining meaning' (2007: 139).

Contexts and meanings[1]

Recent research has identified certain symbols children use to represent highly personal meanings, while at other times the symbols they use are widely understood and accepted by their peers. Graphical marks and symbols provide different 'affordances' or qualities (Gibson 1979, cited in van Leeuwen 2005: 4). For example, a zig-zag mark can suggest its semiotic potential as a symbol to signify stairs (as in Figure 4.1), 'sharp teeth', 'many teeth' or a 'fierce animal'. Kress proposes that children use 'those forms for the expression of their meaning which best suggest or carry the meaning, and they do so in any medium in which they make signs' (1997: 12). At other times children draw zig-zags for another sort of 'power' – to denote electricity and lightning, showing how they select 'no doubt unconsciously – those characteristics which [they regard] as most important for [them] in the thing [they want] to represent (Kress 1997: 93).

In Figure 4.1, Romy explored her idea of a staircase leading down from a great height. The stairs (*inside* her real house) appear to be on the exterior of the house and unconnected to the ground, suggesting what Luquet refers to as 'intellectual realism', where children represent 'abstract elements which only exist in the mind of the artist . . . the essential elements of the represented object, and to preserve each in its characteristic shape' (1991: 102–5).

At school in the Netherlands, Sterre (5 years, 4 months) pointed to the triangles (zig-zags) at the top of her page, explaining that they were 'flags' and that the lower edge of the paper she had cut with pinking shears was the 'shadow of the flags' (Figure 4.2). Other zig-zags in her drawing represented water and 'beach shoes', and some resembling a letter 'M' signified birds in flight.

In another school in the Netherlands, Aman used a twig to draw 'boats' in the sand outside. She completed the top of each with a wavy line, explaining this was 'water' (making a 'hybrid' symbol): it appeared that Aman's intended meaning was *boat-on-water*. Combining and transforming symbols allows children to create and communicate complex meanings (Kress 1997; Pahl 1999a). Pahl (1999a: 20–1) writes of a child transforming a junk model she had made:

> The things that are linked in the mind have become linked in the material world . . . using one idea the children are driven by internal links within them to explore other possibilities. This reflects both the children's inner thoughts and their interest in how the object looks. Both impulses are at work. If an object reminds

---

[1] This section (with adaptations) first appeared in Worthington, M. (2009) Fish in the water of culture: signs and symbols in young children's drawing, *Psychology of Education Review*, 33(1) (March).

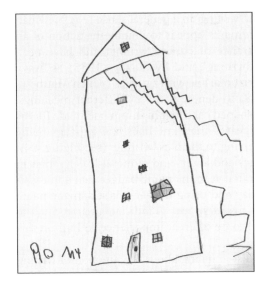

Figure 4.1 Romy's stairs

Figure 4.2 Sterre drew herself at the beach

Figure 4.3 Aman's 'boat-water' symbol

children of something else, they are able to develop it structurally so that it becomes the thing inside their heads. The meanings change and grow inside their minds ... These meanings then develop as they move from one concept to another.

Children also use zig-zags to represent 'writing' in a general sense (e.g. Newman 1984; Kress 1997; Matthews 2003). Such marks appear to imitate the action of an adult's hand as they write, or the appearance of cursive writing, and have been described by Clay (1975) as 'mock handwriting', and by Sulzby (1985) as 'wavy scribbles'. For example, Nathan drew a horizontal line with zig-zags (which Matthews describes as 'travelling zigzags') (1999: 38) to denote two very different meanings: firstly as his 'birthday cake', a 'caterpillar'-shaped cake that his mum had made for his fourth birthday (not shown). He then repeated the same lines and zig-zags many times on the reverse of his paper, now referring to them as 'writing' (see Figure 4.4), whereas Eva used a combination of zig-zags and letter- and numeral-like symbols in her writing (see Figure 4.5). Nathan had often seen the name badges that adult visitors to his nursery wore and on another day and using the same labels, made name badges for his friends and his teacher. His marks appear to indicate a name, and the function and role of name badges seemed to be of great importance to him on this occasion (see Figure 4.6).

Aimee-Rose (4 years, 3 months) was also thinking about the function of writing (see Figure 4.7). Her interest grew from authentic contexts in her nursery in which writing is a common social and cultural activity. Aimee-Rose explained to her teacher, 'I was writing what people were saying, like you do.'

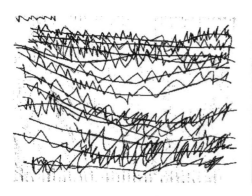

Figure 4.4 Nathan's 'writing'

Figure 4.5 Eva's 'writing'

Figure 4.6 Nathan's name badge

Figure 4.7 Aimee-Rose '. . . like you do'   **Figure 4.8** Lewis's world with planes

By exploring and encoding meanings in a range of contexts, children come to learn how graphicacy can be used flexibly to carry different meanings. Lancaster argues that another significant representational principle is that 'mark-types do not have fixed referents, but can be repeated and used in different environments, the environment being the significant factor' (2007: 139). This can be seen in the examples of zig-zags above, and in the children's use of crosses, below. Crosses also have high visual impact and are often seen in western societies, outdoors on ambulances and road signs, and indoors on remote-controls. The 'x' that signifies a 'kiss' on a birthday card is likely to be the first sign to which children's attention is drawn and which they make. However, young children go beyond these cultural conventions to use crosses to signify many other meanings within the texts they create. In Figure 4.8, Lewis used two different signs to signify one meaning: a cross for 'a plane flying above the world' and in contrast (lower left), an aeroplane in profile.

Meanings are often similarly understood and used by children in different settings and cultures – for example, Nadieh and Kyran both used crosses to represent hands, yet Nadieh attends school in the Netherlands and Kyran attends a children's centre in England (see Figures 4.9, 4.10). Observing the drawing he'd done of his mummy, Kyran commented, 'She's got funny hands!' In Figure 3.11 Nathan made repeated crosses to signify items on his list (as in Clay's 1975 'recurring principle'). His 'writing' there is distinct from that in Figure 4.4, even though they were done within a few weeks of each other.

In their peer groups, signs and their meanings 'move' between children (and adults) and as they appropriate and adapt those they see others use. Tomasello argues that the human ability of *intentionality* (1999: 6) supports sign-making and use. Once children understand that others have intentions 'like me', a 'whole new world' begins to open, 'a world populated by material and symbolic artefacts and social practices that members of their culture, both past and present, have created . . . Children now come to comprehend how "we" use artefacts and practices of our culture – what they are "for"' (1999: 91).

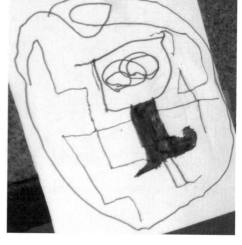

Figure 4.9 Nadieh 'at the beach'

Figure 4.10 Kyran's drawing of mummy

In their *mathematical graphics* children often introduce personal signs such as hands or arrows. These are pointing gestures used either to emphasize the quantity added or subtracted, or to signify operations such as 'subtract' (where a hand is drawn with a quantity of items or a numeral to show how many are subtracted). Children's signs become cultural tools that are available to the peer group and are sometimes adopted and adapted by others (Carruthers and Worthington 2006, 2008). Tomasello explains this as 'cumulative cultural evolution of cultural artefacts' (1999: 40). These shared signs suggest more than one voice or *mutivoicedness* that reflects the speech of others (Bakhtin 1986; Wertsch 1991).

The variety of graphical signs that children use to communicate meanings, and their ability to use them so flexibly, is astonishing. In the 'garage' role-play area outside his nursery, Mark was playing in an area enclosed on three sides. He objected to several boys who repeatedly rode bikes into 'his' corner and insisted 'No! Keep out! You can't come here!' Mark chose an alternative way to communicate his message and, fetching a stick of chalk he drew large crosses (see Figure 4.11), emphasizing his message by repeating his verbal instructions as he drew. Finally the boys 'heard' his request and moved away from where Mark wanted to play (Carruthers and Worthington 2006).

In contrast Daniel (Figure 4.12) had been playing shops and decided to make a sign to show when the shop was 'open' and another to show that it was 'closed'. His teacher had noticed what he was doing and Daniel explained:

Daniel: It's closed now, the café is closed.
Teacher: How do I know it's closed?
Daniel: [points to his picture of a face crossed out] Look here, see? Closed, that means it's closed.

Figure 4.11 Mark: 'No! Keep out!'
(4 years, 0 months)

Figure 4.12 Daniel's 'shop closed' sign
(5 years, 1 month)

Then he rubbed out the drawing and drew a face without a cross, explaining, 'Look! Open that means it's open now . . . Oh dear.' Drawing a cross over his drawing of a face, he explained, 'It's closed.'

Children's use of graphical signs shows that sometimes they use a sign as a *prototype*, such as the use of zig-zags to signify 'writing', however, this often relies on some shared understanding since without it such symbols would be meaningless.

Flexible uses of graphical symbols

Drawing allows free-ranging explorations of graphical signs which support semiotic development and deep understanding of the way in which graphical signs can be used to signify particular meanings in various social contexts. Matthews argues that children's drawing 'is not defined in terms of a body of knowledge . . . and simply transmitted to the learner. Nor is it tied to the transmission of any particular culture'. He encourages teachers to 'understand some of the mechanisms which drive representational thought' (1999: 163) and raises concerns about young children's experiences of drawing in educational settings that are shared by others (e.g. Anning and Ring 2004; Ring 2005).

However, although Matthews (1999) proposes that writing may emerge from drawing, we believe there are problems with this perspective, not only because it may suggest to teachers that drawing is 'useful' only as a precursor for writing but because – from our evidence – drawings, writing, maps and *children's mathematical graphics* appear to develop *simultaneously*. This perspective is supported by Karmiloff-Smith (1992) and Lancaster (2003).

Children's use of graphical symbols for mathematics

In addition to numerals, children use various symbols to signify personal mathematical meanings. In the next example (Figure 4.13) Giorgio used zig-zags to mean 'peoples': this appears to work in a similar, generalized way to children's use of zig-zags for writing. Giorgio lives in Rome but attends school in Zurich, and each weekend he travels home to Rome, a long journey for a 4-year-old. Giorgio was exploring his feelings about returning home combined with his passion for trains. At the same time he was expressing his thoughts about 'lots of people': his personal use of zig-zag symbols was sufficient for him to signify an *uncounted quantity* of 'peoples'.

Tommy's class had been on a visit to the zoo and were arguing about which was the 'best' animal. Tommy (4 years, 7 months) decided to ask his friends and after he'd drawn a lion, a crocodile, a giraffe and a zebra, took his clipboard to collect his data. Figure 4.14 shows how he decided to use crosses to indicate which animal each of his friends liked best (representing quantity of children he had counted). The choice of crosses to denote each child's choice was his idea. In Figure 4.15 Anna used crosses for calculations (using the standard addition sign she knew), to calculate the combined totals on the two dice she threw in a game. These two examples also provide a glimpse of one of the challenges of sign use in mathematics. Whereas in written English 'x' stands for a letter of the alphabet and its related sound, in mathematics (as in drawings) one sign can mean very different things and is highly dependent on its context.

Figure 4.13 Giorgio's 'peoples'

Figure 4.14 Tommy: 'What's your best animal at the zoo?'

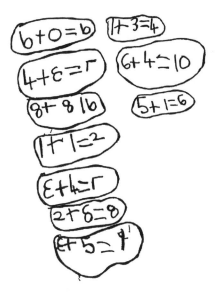

Figure 4.15 Anna's dice game

Discussions and definitions of 'mark-making'

'Mark-making' is increasingly used as a generic term to describe the ways in which children 'create and experiment with symbols and marks', for both drawing and writing (DfES 2007a: 64). However, there is considerable confusion about the meaning of this often undefined term, and our concern is that it lacks clarity, failing to do justice to young children's powerful thinking and the ways in which they choose to explore and communicate that thinking.

We are not alone in raising these concerns about the language used to describe children's visual representations, and in early childhood research the term 'graphics' and 'graphicacy' are gaining ground (e.g. Anning 2003, 2004). For example in a presidential address to the National Society of Art Education, Gallon argued, 'Too often the glib and trite definition of "drawing" is given as "the making of marks"' (1981: 3). Ring (2001: 1) argues that: 'Also damaging to some extent, for the understanding of the role of drawing in young children's learning, has been the exchange of the word "drawing" for "mark making" in educare settings.' Ring (2005: 2) proposes that:

> The term, in emphasising the importance of children's earliest marks for writing development, can give the message that pictorial representation is inferior to the more important role that the reading and writing of symbols has been given within the National Curriculum and within society in general. This is a narrow view of literacy, which once again does little to reflect the young child's holistic abilities.

Citing Ring (2001), Hall observes, 'Drawing is more complex than mere "mark making", and although this is a popular term in early years circles it actually undermines the importance of drawing in early years education' (2007: 3–4).

We realized early in our research that to term young children's mathematical representations as 'mathematical *marks*' was insufficient since it ignores their use of drawings, personal signs in drawing, early writing, letters and words, tally-type marks, personal and standard mathematical symbols and the complex ways in which they arrange and combine these within their mathematical texts such as calculations or data representation.

Conclusion

This chapter has highlighted the meanings and contexts of children's graphics, as young children use graphical symbols in significant and complex ways. In the next chapter we explore the graphical representations of the very youngest children, from birth to 3.

Reflections

- As you observe children drawing, writing or engaged in other graphics, look out for any specific symbols such as zig-zags or crosses that you see them use. You are likely also to see other symbols (e.g. hearts or arrows).
- With your colleagues, look at several examples that include the same symbol: what did the child mean by using that particular symbol? Are their meanings the same?
- During the course of the term, collect all the examples you see where children have used one of these symbols (e.g. arrows) and discuss their meanings and purposes, and the contexts in which they were used.

Recommended reading

Athey, C. (2007) *Extending Thought in Young Children*. London: Paul Chapman.
Clay, M. (1975) *What Did I Write?* London: Heinemann.
Cox, M. (2005) *The Pictorial World of the Child*. Cambridge: Cambridge University Press.

5

Babies, toddlers and two-year-olds

'Babies are not just as gifted as adults, they are much more so.'

(Mehler and Dupoux 1994: 168, original emphasis)

What is this chapter about?

- **Very young children's capability to represent.**
- **Scribbles.**
- **Symbolic meanings.**
- **Young children's marks as systematic investigation.**
- **Children under three thinking mathematically.**
- **Pegagogy.**
- **Gender and children's graphics.**
- **Thinking beyond 'messy' play**

This chapter is about children's graphics from birth to three. It is important to separate out this age group as the abilities of children under three are often underestimated and sometimes ignored. These young children's experiences and how they are valued is sometimes only at the care level, especially in institutions. Often they are well looked after and loved but their intellectual aptitude is not always understood and therefore not responded to. And yet nurturing children's intellectual capacity when they are babies, toddlers and two-year-olds provides a strong base for their educational life (Gopnik *et al.* 1999).

Very young children's capability to represent

Baby humans have the ability to develop into persons who can represent ideas in many ways. These young children, if given the opportunity, choose to use different kinds of media in the ways they desire. Babies and toddlers experience the world through exploration and experimentation, and this is also how they develop graphicacy.

Through a constant drive of curiosity they begin to represent their thinking. Matthews (2003) has observed babies and very young children painting and drawing, and explains that this is the beginning of cognitive development, visual representation and expression. Babies are born with the ability to think (Gopnik *et al.* 1999), and as they use materials to experiment with line and form they also are representing their thinking. Babies previously were not seen as thinkers and the acknowledgement of them as such is extremely important because the adults around them will take their actions more seriously. A recent project showed that practitioners in baby rooms were surprised that babies knew as much as they do, and revealed that babies understand far more than we think they do (Lawrence *et al.* 2008). Before they are one, children are manipulating and playing not only with objects but also with ideas. In relation to babies and toddlers, Matthews (2003) emphasizes that abstract thinking begins in dynamic play and exploration.

Scribbles

Luquet's seminal research in 1927 professed five main drawing stages. The first was from birth to two and a half years. In this 'scribbling stage' Luquet remarked that the child makes marks which are not intended to represent anything. At this stage the mark-making is a sensory motor investigation, and is enjoyable (Luquet 1991). This links with Piaget's (1958) 'sensory motor stage' and Burt's (1921) 'meaningless scribbles' stage. However, these 'stage theorists' do not take into consideration the sociocultural aspects of the child's world and their interpretations are based on the goal of successful drawings from an *adult perspective*. More recent research has opened up this debate and Matthews (2003) makes a strong case that artistic development is socioculturally based. He strongly criticizes the view that visual realism is the aim of children's drawing and paintings. Graphics by children under three are complex and much debated, and there is an overwhelming case that they are not meaningless scribbles. Malchiodi (1998) recognizes that a child's first scribbles symbolize a 'developmental landmark' since they are now able to make connections to the world around them with their marks on paper.

Despite this, the tendency to think of children under three using materials in a random and non-intellectual way still persists. Lancaster (2007) suggests that many would consider it unreasonable to even research whether children under three construct their own meanings and signs, as if it was ludicrous to even think about such a concept.

It is difficult to research the representational capacity of young children. Babies and children under three present not only practical difficulties but are still developing language, and this may place limitations on the adult to comprehend exactly what they are saying and to provide explanations. The child's 'key person' in the nursery (Elfer *et al.* 2003) is more in tune with them, but might also miss some vital aspects of their graphical meanings, whereas parents are much more at one with their own children and pick up nuances that others would miss: they know what the children's interests are and who are the important people in their lives.

When studying representations from children under three, Lancaster claimed an advantage over other researchers due to the up-to-date technologies that were used (2007). The emphasis of her research was on interpreting meanings rather than simply describing and reporting, and this was a significant factor in uncovering the capabilities of this age group. From this study Lancaster argued that whatever their age, children's marks hold representational significance for them, and that any difficulty that exists lies in identifying the children's intentions. The main finding of this study was that children under three have the capacity to use graphic marks with intention and meaning.

Symbolic meanings

According to Lancaster, 'Children have a social and personal investment in the symbolic meanings of their culture right from the start, and it follows this engagement with them is likely to be always intentional and purposeful' (2007: 132). Many graphics created by young children can be misrepresented as 'drawings' – in other words, the child is 'drawing a picture'. The word 'drawing' tends to be used generically to describe all young children's graphics. In nurseries and preschools, adults working with children also use the term 'picture' and phrases such as 'drawing a picture' as general terms for all children's graphics. Most see the aim of children's actions on paper as to represent something that is an exact likeness of an object or place in the real world: we may call this the 'pretty picture' syndrome. Some children are even judged on their ability to draw a person (Goodenough 1926). There is also doubt that children have the physical ability to draw and that any encouragement of this is 'not right' (David et al. 2002). We would argue that children do not need a highly developed physical ability to make meaning in their graphics.

Research of very young children's graphics reveals that many of these early creations (those that they choose to make) are *not* 'drawings' in the sense of pictures or direct representations – for example, 'drawing' a fish or a man – but are children's own ways of representing their thinking through symbols (Cox 2005). Symbols are abstract forms of representation. For example, a child might draw a long line for a man or a circle for a car, *intentionally*. The child is deliberately using their own symbols to represent their own concept of what a man or a car is, not necessarily because they lack the skills to represent the exact image of a man or a car. This symbolic representation is important as it forms the basis of the symbolic languages such as mathematics and writing. It is vital to acknowledge that children even as young as one are symbol users. The nurturing of the child as a symbol user is important because it is essential to the child's own understanding of the abstract symbolism of mathematics. However, this does not mean that we expect very young children to sit down and make symbols in line with adult objectives or government targets. Vygotsky (1930) suggested that children under three learn while following their own agenda; therefore, the environments we provide and the knowledge and interactions of the adults working with the children are crucial.

Bruce (2004: 170) emphasizes that 'Children take great joy in becoming symbol users, providing their personal symbols are valued, and provided they are not forced into early and inappropriate adult-led tasks'. It is important not to lose the child's meaning because that is how they understand the world. Set activities that do not take into account the child's thinking become tiring for young children and squeeze out their own thinking as they desperately try to fit in with the adult's outcome. It is important to promote children's natural ability to form and make meaning with symbols.

Young children's marks as systematic investigation

Young children's early drawings are complex; they are part of a 'systematic investigation, rather than haphazard action' (Matthews 1999). Work by Matthews with young children making marks (graphics) revealed that they are exploring line and form, and he categorized their work into three 'generations':

- The first generation of marks are horizontal and vertical marks, dots, blobs, spots and 'push pull' actions. Children learn to separate and recombine drawing actions in a variety of ways.
- The second generation includes continuous rotation, demarcated line endings, travelling zig-zags and continuous lines.
- The third generation builds on the first two and involves closure, inside and outside, core and radial, parallelism, collinearity, angular attachments, right-angle structures and u-shapes on a base line.

'Closure' is a significant structure because it gives the children more possibilities as they are now able to separate 'outside' from 'inside'. Children in their investigations will vary the size of line, closure or zig-zag. They will vary how they use the paper and sometimes it will become of itself a dynamic part of their intention. For example, after drawing on the paper children will sometimes cut it and use it in their imaginary play. They make meaning in a plethora of ways with a plethora of means, in two, three and four dimensions (Kress 1997). Matthews' work is supported by Athey (2007), whose study revealed that children have schematic interests, and often these are seen in their graphics (2007). For example, a child who is interested in circles or rotation will produce graphics that have these aspects to them.

It is easy for adults to discard such 'scribbles' as meaningless, however they are precious records of children's thought processes. When adults are in tune with children, the scribble-like marks will be valued, carefully annotated, discussed with parents and placed in the child's record to show their development and thinking.

As a baby, Emma loved to explore the paint and brushes in the nursery (see Figure 5.1). Exploring different kinds of media seemed to be Emma's preference for a long time and she developed into a very prolific explorer of graphics. Rosa discovered that cornflour is sticky and messy (see Figure 5.2). She played with it and found that she could make a mark that covered up very quickly and vanished. She did not convey her experience to the practitioner – she was totally involved in it and had not yet developed the words to express her experience. Her gesture and facial expressions show her total involvement. Her new experience was gently encouraged by the practitioner. Rosa's manipulation of this material was powerful and she was in control of what to do and where to take this experience.

Opportunities for children to experiment with all kinds of materials in their own ways are important. Such materials include paper and pencils. A toddler will not

Figure 5.1 Emma (9 months)

Figure 5.2 Rosa (13 months)

Figure 5.3 Finn (1 year, 8 months) is keen to make his own graphics on the road tracks. Susan, his key worker, is in the background gently talking to him

necessarily use the paper or the pencils in the conventional manner but may explore them by stabbing at the paper, tearing or folding it. These experiments are not necessarily a precursor to writing or written maths, but they are a part of the whole experience and as children become older they will still be experimenting in this way. The important point is that materials should be made available for children to explore and that teachers and practitioners understand how important this is. Sensitive interactions start with sitting and observing the child as they explores the media and their thinking. For very young children, who may not be totally physically active, the practitioner may act as an anticipator of need, in the background, gently waiting for the child to lead (see Figure 5.3).

Children under three thinking mathematically

From birth, as well as having the ability to use and experiment with graphical marks and objects in an abundance of ways, children are also developing mathematical thinking. As young as five months babies appear to be aware of quantities and can notice changes in the amounts of objects (Starkey and Cooper 1980). Athey (2007) researched children's thinking through their schemas and revealed that they have patterns of behaviour which seem to indicate their current thinking. Carruthers and Worthington (2006) highlighted that the majority of children's schemas have mathematical aspects. For example, children involved in a containing schema could also be exploring the space and shape of a box, looking at angles, corners and sides. A study by Carruthers (2007a) traced the development of a group of children's mathematical thinking in an under three's nursery at a children's centre in England. These children were engaged in many mathematical enquiries relating to cause and effect, space and shape, measurement, area, capacity, length, distance, angles and lines. They were in a play environment following their own lines of enquiry with open and natural materials which they used in different ways. It has been shown that the way children use resources often predicts their learning potential (Nutbrown 1994).

Children as young as one continue to be interested in quantities through, for example, the making of piles or the collection of objects, and this interest can develop, as they begin to talk, into counting objects and saying the numbers. A study of a child from eighteen to thirty-six months (Carruthers 1997a) revealed that she had tuned into many aspects of mathematics, including number, using numbers in personally meaningful, social and cultural contexts. The influence of this child's family and especially her older sister were key to her development.

Children explore many mathematical areas before they are three, using action, gesture and language to demonstrate their mathematical knowledge as they play. They also know that numbers are symbols that can be used in a variety of social and cultural ways. Some children begin to use their own ways to write down numbers. They also use graphic materials to explore their mathematical schematic interests.

Finally, this natural curiosity about mathematics that children investigate, and the ability to explore the use of graphics with their own symbols and by combining objects, all come together. Very young children will merge their knowledge and the tools they choose to make new meanings. Children gradually combine two meaning systems together, in a similar way to Vygotsky's interpretation of children's language development (John-Steiner 1985). This is crucial to their growing knowledge of symbols and mathematics.

Pedagogy

There is no doubt that many adults are keen for children to draw something that looks 'like' something in the world, although many graphics by children under three may not look like anything at all to an adult. Children sometimes say their graphics are 'something' and they become 'a something' to the child. Perhaps it is more important to understand these scribble-like marks and not expect them to be anything that has adult sense, instead looking at them from a child's perspective (Carruthers 1997b). In addition, we as adults may sometimes never uncover the true meaning of a child's graphics but should rather value them just for themselves.

Sometimes, when children are making their own graphics on paper, teachers ask 'What is it?' Some children will look and say something that they did not intend, while other children will look blank because this is a puzzling question, especially when you were not drawing 'a something'. For many children the curves, lines and circles that they are producing are more than enough in themselves; they are exploring zig-zags, circles, or dots, dabs or combinations of these. The graphics children produce are intentional even though to an adult they may look haphazard. Children will be thinking something that they may not verbally be able to express.

It is such a sensitive matter to respond appropriately to a child's scribble-like marks, and Matthews (2003) suggests that teachers need to properly understand children's drawings to give a genuine response. This is part of the pedagogy – to 'know' about children's drawings. Many of the symbolic ideas children use are not elaborate – for example, as suggested earlier, a line can represent a person. Cox (2005) studied a nursery class for a year and described the graphics of one child who had said her drawing was a boat with her mummy and daddy inside. On further reflection she said the boat was very rocky so she had better attach them together. Rather than redrawing, she chose an economical symbolic solution, drawing some dashes to join them up.

Young children do not necessarily have a fixed goal in mind when they are using materials such as paint, pencils, charcoal and junk: they go with the flow, and their planning is of the moment, and flexible. This does not imply that they never have a goal in mind, but that they may change their original thought as they work with the materials. This is of vital significance to pedagogical understanding. It separates the teachers and practitioners who are 'pretty picture' seekers and those who seek the intellectual curiosity in young children's graphics and strive to understand the children's meanings.

Matthews (2003) goes on to say that many young children will be aware of painting and drawing media but the way they are introduced to these is important and affects their development. If children are given an object or subject to draw, they will oblige, but when they are given opportunities to choose the time, the media and their own thinking then this has huge benefits for their own understanding.

CASE STUDY

Changing practice

Staff at an inner-city children's centre had been developing their understanding of children's graphicacy for four years, and hosted an exhibition of children's graphics in 2007 and again in 2009. The annotations the teachers and practitioners made were comparatively different from one exhibition to the other. In the under 3s nursery the annotations in 2007 mainly consisted of comments about what hand the child had used, pencil control and colours. In 2009 the annotations described the children's understanding, meanings and connections. This was a significant shift in pedagogical understanding and awareness of children's graphics. What made the difference?

The culture

The culture shifted from seeking to find fine motor control and adults' views of pictures, to valuing and understanding the meanings of each child's graphics.

Professional development

Professional development within and outside the centre supported the adults' understanding of the complexities and significance of children's graphics. It was key that the staff had up-to-date professional development that challenged their thinking.

Time to talk

It was important to arrange the timetable to give all the staff in the under 3s nursery time to talk about the children individually: each member of staff had a mentor.

Everyday practice

Listening to the children enabled the staff to bring all they understood to the annotations. When the staff appreciated that there was more to know, then they wanted to know more.

Gender and children's graphics

The difference in levels of achievement, especially in writing, between boys and girls, is an issue that has generated a great deal of concern and debate for many years. Boys seem to thrive in play environments:

- when they combine tools and materials (e.g. scissors, glue, sticks);
- when they are given opportunities to make large physical movements;
- when there is a choice of mark-making implements;
- when we consider their interests and plan from them.

For example, Rio was very interested in cars and played with small vehicles inside the nursery and the child-sized car outside (see Figures 5.4, 5.5). He engaged in graphics

Figure 5.4 Rio (2 years, 2 months) writes parking tickets

Figure 5.5 Rio gives himself a parking ticket

and used white stickers for parking tickets on his car. These choices appeared to be related to boys' interest in vehicles and technology.

Thinking beyond 'messy' play

'Messy play' in an under three's environment can be therapeutic and enjoyable. Children experiment and sometimes use graphics. If, as we have argued, these graphics are intentional and not just haphazard, then as teachers we need to be looking more closely and responding appropriately to the messy play. There is also a question about the definition of mark-making (see page 51). Is printing mark-making? Is spilt juice mark-making? Children enjoy exploring all kinds of tactile and malleable materials, but focusing on the form of their lines, dots and enclosures can help adults understand their intentions and meanings. If we look at messy play *only* as enjoyable and fun then we will fail to take young children's meanings and graphicacy seriously. Unfortunately, many adults continue to regard messy play as children at the sensory-motor stage, rather than thinking about or understanding their actions. Piaget's sensory-motor stage has given credence to messy and enjoyable interactions with materials, but unfortunately has also detracted attention from looking at children's meanings, and what they produce is often used as decoration and display items, or discarded, rather than analysed to uncover personal intentions.

From birth onwards, children are learning about how the world works and so they are looking at adults, other children and the environment to see what systems, such as writing and mathematics, are used. When they commit themselves to putting something on paper, card or any surface it is with their knowledge and their own intention. However, at a very young age their graphics are not underdeveloped forms of any conventional system like writing, drawing or mathematical notation, but are always logical and purposeful *in their own right*.

In a classic example of 'messy play', the children shown in Figures 5.6 and 5.7 are using different tubes of colour to explore lines and form. The practitioner, Lisa, is writing observations of what the children do and say.

Figure 5.6 exploring with paint, focused on circular marks

Figure 5.7 Amber (2 years, 3 months) squeezes the bottle, making dripped dots

Each child's graphical journey is unique; children respond differently because their cultural and social experiences are different.

CASE STUDY

Yasin and Spiderman

Child

Yasin

Teacher

Carole

Practitioner and team leader

Clare

Yasin (1 year, 11 months) awoke from his sleep: he was cuddled and supported back into joining the afternoon nursery session. He indicated he would like to go outside and once there he gestured to the adult that he wanted her to share a book. The two of them settled into a large beanbag with the book Yasin had selected. When they had finished with the book, the adult asked Yasin if he would like to read anything else and he nodded his head and said, 'Spiderman.'

Yasin fetched the Spiderman comic and settled back into the beanbag. The adult responded to his directions in turning the pages. He pointed to the details in the pictures and made comments. He then spotted the template of Spiderman in the comic and rubbed this with his finger. The adult asked him if he would like a pen to draw on the page and he nodded.

The adult fetched some pens and Yasin sat at a table and began to draw. It was at this point Yasin looked over and noticed some chalk marks on the pavement which he went over to examine. The adult found some chalk and offered it to Yasin who began to make his own marks on the pavement. He labelled his marks, 'Spiderman' (see Figure 5.8).

Yasin then made other marks on different surfaces outside; on the planks of wood and the steel pole (see Figures 5.8–11). As he did so, he repeated, 'Spiderman.' He then pointed to a set of marks he'd made on the playground and said: 'Mummy's.'

At snack time, Yasin was still repeating, 'Spiderman.' Carole tied some wool loosely around Yasin's wrist to emulate Spiderman. He sat holding his hand out saying, 'Spiderman – web.' The other three children at the table requested a piece of wool each on their wrists and also set about 'spinning webs'. While enjoying their snack, the children looked at the images in the Spiderman comic and continued to spin webs. Looking through the comic, Yasin spotted a picture of the Manchester United football team, pointed to it and said, 'Rooney.'

Figure 5.8 Yasin (1 year, 11 months) draws on the pavement

Figure 5.9 Yasin draws on the wood

Figure 5.10 Yasin draws 'Spiderman'

Figure 5.11 Yasin continues to draw

At the end of snack time, Clare asked Yasin, 'What would Spiderman like to play with?' and he replied 'Spiderman car.' Carole lined up the chairs and Clare went over and sat down with the children. She supported the children in finding suitable 'steering wheels' and initiated singing Spiderman car songs. Yasin then said the car was a train and Clare led a 'Spiderman train' song. Other children returned from a visit to a farm and the Spiderman play petered out.

Carole's reflection on what this tells us about Yasin's learning and development

'Yasin is interested in print and illustrations and definitely in Spiderman. He is familiar with aspects of popular culture, which reflect his wider home life experiences, for example the footballer "Rooney". Yasin makes marks to represent. He has energy and focus, making marks in a rapid fashion all across the outside area on various surfaces.'

Over the next few weeks Carole provided further learning experiences for Yasin around his interests in super-hero play and Manchester United which both included opportunities for graphicacy and reading similarly themed texts.

Prolific mark-makers

Some children are prolific mark-makers, producing many specimens based on their graphic-making interests. Other children have long periods of time where they have a heightened interest in making their own graphics but then stop. Most children from six months to two years are deeply interested in experimenting with any materials that are accessible to them and take all the opportunities given to them to wallow in these materials. Some become very intent on these experiences.

CASE STUDY

Maps of the world

Practitioners

Louise and Maria

Some children are fascinated with map-making and maps (see also Chapter 3). The practitioner's role is to support children's interests and provide the materials to stimulate further conversations.

Macey and Orna examined a map of the world (see Figures 5.12, 5.13) after they had asked Louise about Misaki (a practitioner who had returned home to Tokyo) and where she had gone. This stimulated interest in maps and map-making.

Tiyanni was in the garden, sitting at a table with Maria and Macey. Together they were looking at a map in a large diary. Maria talked about the time Tiyanni had gone to Jamaica and pointed to Jamaica on the map. Tiyanni explained, 'That's where my grandma and grandad is,' and Maria talked about the beach in Jamaica. Picking up a blue pen Tiyanni made some marks in the diary and, pointing to them, added, 'This is swimming' (see Figure 5.14).

Macey was looking at the map in the diary and Maria pointed to Tokyo and explained, 'That's where Misaki comes from.' Macey marked a dot on the map where Tokyo was and as Maria continued to talk about other countries, she added dots to mark them on the map (see Figure 5.15). Turning to another page in the

Figure 5.12 Louise pointing to England

Figure 5.13 Orna (2 years, 4 months) and Macey, looking for Japan

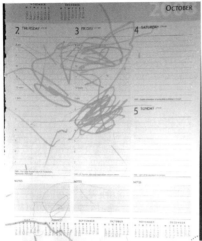

Figure 5.14 Tiyanni (3 years, 2 months)

Figure 5.15 Macey (2 years, 10 months)

Figure 5.16 Macey: 'That's where
I live'

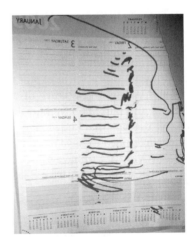

Figure 5.17 'That's where
Tiyanni lives'

diary, Macey made small vertical marks in a column. Turning to yet another page, she made marks beneath 'Tuesday' and said, 'That's where I live' (see Figure 5.16). She then drew another column to represent where Tiyanni lived (see Figure 5.17).

Using Whiteboards and available materials anywhere anytime

Mikey used an action narrative to explain his graphics to Lisa. 'This is Sandy Man. He goes round and round.' Then Mikey jumped round and round (see Figure 5.18).

When resources such as clipboards are everywhere, inside and out, children will take the opportunity to use them in an abundance of ways to suit their purpose (see Figure 5.19).

Figure 5.18 Mikey (2 years, 2 months)

Figure 5.19 Andrew (2 years, 2 months) is taking notes in the forest

Conclusion

There is much evidence that supports the premise that young children under three can and do make their own mathematical representations. They explore media when they are given the opportunities and have intention and meaning in the graphics they produce.

The adults who are the key figures in a child's life have an immense role in nurturing their graphical explorations. This includes provision of a variety of materials and surfaces. A culture of enquiry needs to pervade the setting. It is vital that teachers and parents have knowledge of the ways very young children represent and are able to share that knowledge with each other. Finally, adults working with children under three need to have open minds about the ways that very young children represent. It is vital not to have a predetermined outcome.

In the next chapter we explain *children's mathematical graphics*.

Reflections

- Bring together examples of very young children's self-initiated graphics. Analyse them in terms of context (where and why), lines and form, and in terms of what the children told you about them, if anything.
- How do you talk to the children about their graphics?
- Do you have a range of graphic materials for children under 3 to choose from?
- What are your children interested in?
- Read the Yasin and Spiderman case study again. How have the adults supported his play?

Recommended reading

Paley, V.G. (1986) *Mollie is Three*. Chicago: University of Chicago Press.

6
Children's mathematical graphics

'Christopher, are you three years old?' ... 'I got three on my birthday, that other day,' Christopher says. 'Me too,' Mollie states happily, taking Christopher's hand. 'We're both the same three and so is Frederick. He told me.' ... Mollie is alone with Christopher. 'Here's your birthday cake. It's your party.' She rolls out a cylinder of playdough. 'And here's your older candle. Elephants need big candles.'

(Paley 1986a: 49)

What is this chapter about?

- **The beginnings of mathematical notation**.
- **Symbol use in mathematics**.
- **Research into children's mathematical graphics**.
- **Written mathematics in early childhood education in England**.
- **The taxonomy and dimensions of children's mathematical graphics.**

In the first four chapters we described and illustrated aspects of children's meaning-making in their play. Chapters 1 and 2 introduced children's symbolic play, showing how they use a variety of media and means to represent and communicate their thinking, including their mathematical meanings, in various play contexts. Chapter 3 explored children's graphicacy through their drawings, writing and maps, revealing how culture influences their symbolic representations. Chapter 4 focused on symbols children use within their representations and the range of contexts in which they use them: it highlighted the ways in which children use graphical symbols flexibly to signify and communicate specific meanings. These four chapters uncovered some of the complexity of children's thinking through graphicacy, revealed through many layers and accentuating the holistic nature of young children's learning. Chapter 5 illustrated very young children's capabilities to represent their meanings through their graphical marks and representations. Now, in this chapter, we continue this journey, focusing on children's 'meaning-based' symbol use in mathematics: their *mathematical graphics*.

We explore some of the research into 'written' mathematics and provide an overview of our research into children's mathematical graphics, using the taxonomy to support teachers' understanding.

The beginnings of mathematical notation

In spite of the fact that 'written' mathematics has been shown to cause young children considerable difficulties (e.g. Ginsburg 1977; Hughes 1986; Pimm 1995) there has been very little research into the beginnings and early development of young children's own mathematical representations and it is unsurprising that teachers and practitioners have been unaware of the potential of this aspect.

Vygotsky and Luria's research in the 1930s was the first to reveal that young children could use their own marks to represent meaning (as writing) (Vygotsky 1930, 1983; Luria 1998). Other researchers such as Marie Clay (1975) and Ferreiro and Teberosky (1982) then went on to study young children's early writing, leading to current research on the 'new literacy' (e.g. Pahl and Rowsell 2005, 2006).

Researching young children's mathematics, Ginsburg found that 'Children's understanding of written symbolism generally lags behind their informal arithmetic' (1977: 90), and identified differences between children's informal and formal knowledge of written mathematics. This, he observed, often results in a gap between their intuitive understanding and their ability to use formal abstract mathematical symbols such as written calculations. He argued that 'While they make many errors in written arithmetic, children may in fact possess relatively powerful informal knowledge', and that this knowledge can be used as the basis for effective instruction (1977: 129). Aubrey (1997: 49) concluded 'that a major difficulty for school learning is the lack of account taken of this rich, informal knowledge derived from everyday problem solving situations'.

Carraher et al. (1985) revealed the inventive strategies used by young Brazilian street traders to make computations, compared with the difficulties they had with similar problems in formal school mathematics. In their research, Nunes et al. (1993) emphasized 'oral arithmetic' as the means that allowed such children 'to calculate and to think of the values that they are working with at the same time . . . In contrast, in [standard] written arithmetic we set the meaning of numbers aside during calculation'. According to Nunes and Bryant (1996: 107), 'This seems to detach the children from the meaning of what they are trying to calculate'.

Hughes was the first to show that young children could use their own marks and symbols to represent small quantities that they had counted. He proposed that images have a significant role in 'translating' between 'different modes of representation' (1986: 170), their own intuitive and informal representations and formal 'school' mathematics. Others have added to this argument that children's informal understandings of 'written' mathematics should be valued (e.g. Peterson et al. 1989; Gifford 1990; Fuson et al. 1997; Brizuela 2006). In England, informal written methods and a need to strengthen the relationship between mental and written mathematics is increasingly emphasized in official reports, including the annual subject reports from the Office for Standards in Education (Ofsted) (e.g. 2002, 2008), although we have a way to go before this is fully achieved.

Symbol use in mathematics

Many researchers have explored the significance of language and symbol use in mathematics (e.g. Ernest 1991; Cobb *et al.* 1992; Pimm 1995) and acknowledge the significance of children's own representations. For example, Terwel *et al.* (2009: 26–7) explain that 'in designing representations, students learn how structuring processes develop . . . This knowledge enhances the capacity to generate new solution processes and to transform a representation according to changes in the situation, facilitating the construction of solutions to relatively new and unfamiliar problems'. Representations also 'play a major role in problem solving'. They propose that firstly, graphical representation 'closes the gap between prior knowledge and the material [children] are involved with. Second, it provides opportunities for creative engagement and ownership of conceptually difficult material. Third, it enables students to exercise their 'meta-representational knowledge', which is expected to be of value in the creation of new representations (2009: 28–9). Van Oers argues that:

> One of the core concepts in all dynamic views on mathematics is the concept of symbol. More importantly, however, symbols also function as means for regulation of the thinking process. They introduce new ways of organising in the course of thinking, as argued by Vygotsky and his followers . . . Representations that retain much of the details of the original experience are termed perception-based representations. Their counterparts are *meaning-based representations* that represent the phenomenon or situation from a particular point of view and discard unimportant details.
>
> (van Oers 2001, in Terwel *et al.* 2009: 27, emphasis added)

Children's mathematical graphics are cognitive representations, underpinned by cultural-historical and semiotic theories (Vygotsky 1978; Kress 1997). 'The invention, use and improvement of appropriate symbols', van Oers emphasizes 'are the characteristic features of the mathematical enterprise . . . the efforts of the pupils to get a better grip on symbols in a meaningful way should be considered one of the core objectives of education, especially in the domain of mathematics' (2000: 136). Pape and Tchoshanov (2001:126) argue that the 'educational significance' of such a 'holistic approach' is 'characterised by its completeness and orientation toward creativity though understanding'. In revealing their thinking, children can also communicate their mathematical ideas, adding to the cultural 'pool' of semiotic tools that can be appropriated, used and adapted by others in the social group. Terwel *et al.* (2009: 41–2) propose that such representations:

> may develop a sense of knowledge ownership, which makes students feel free to transform this knowledge as the situation requires . . . children who learn to design are in a better position to understand pictures, graphs, schemes, models, or similar intellectual tools and are more successful in solving new, complex mathematical problems on relations and proportions. In short, the process of designing representations in collaboration is a way of reinventing the more abstract mathematical structures and principles. The process of reinvention, in

turn, enhances problem solving in new situations. It follows that increased attention to the process of model designing may be useful in classroom practice.

The case studies in this and in the following three chapters focus specifically on *exploring and communicating mathematical thinking* through graphical representations in a range of contexts. The following study provides an interesting insight into symbol use in contexts that are neither specifically 'drawing,' 'writing' or mathematics.

CASE STUDY

Scoring goals: sign-making in children's self-initiated ball game

Children

Henry (4 years, 4 months) Joe and Thomas (both 4 years, 7 months)

Teacher

Mandy

Context

The boys were using the basketball net and talking about how many goals they were scoring when Mandy suggested that they might like to use the chalk to 'put something down'

Dimension of the taxonomy[1]

Written number and quantities: *exploring symbols; representing quantities that are counted*

After some time Henry picked up the chalk and began drawing symbols on the ground: the other boys joined him and soon their symbols spread over the playground. During the course of their play they frequently called Mandy to look and explain what their symbols signified (see Figures 6.1–6.6).

Figure 6.1 'A cross means you lose'

Figure 6.2 'It's a *double* lose! This is where you *double-lose!*'

[1] For Dimensions of the taxonomy, please see pages 75–7.

Figure 6.3 'A cross means you *win* – a straight line means you lose'

Figure 6.4 'It's such a long way; it means you double-win the whole match! Come on keep going.' 'Look how long we've drawn!'

Figure 6.5 'You win here, you win here, you lose here'

Figure 6.6 'A circle is six goals. I got six goals – I landed in the circle!'

The boys played this game for over an hour, stopping only because it was time to tidy up and go home. In the following days Mandy created an outside display of the boys' game and the complex symbols they had created (Figure 6.7).

Figure 6.7 The nursery's outside display

This symbolic flexibility enables children to also explore and communicate symbols in mathematical contexts, and negotiating and co-constructing meanings through dialogue ensures that their signs are deeply meaningful.

Research into children's mathematical graphics

Our original classroom-based research begun during the 1990s when we were teaching in nursery, Reception and Key Stage 1. We had questioned the 'written' maths that the children did and felt that there must be something more effective. Comparison with the children's emergent writing at that time revealed a marked difference between the two. By choice the children we taught often elected to write spontaneously and were confident writers, the content of their writing was often rich and diverse: it was rooted in their own experiences and also showed a depth of understanding of spellings and punctuation. As teachers we had both developed our understanding of the theories underpinning emergent writing and the pedagogy to support it, and this was our starting point for developing effective 'written' maths. Above all we believed that if children could build on their informal understandings of using marks, symbols and other representations, this could enrich their understanding of the abstract written language of mathematics. At that time we had not envisaged the extent to which an 'emergent' approach to mathematics would support their mathematical thinking at such a deep level.

The children began to use their *own* graphical representations: they employed their chosen graphics – scribble marks, informal and standard symbols, drawings, letters and words in texts that were full of *their* meanings (rather than ours). As we observed the children and collected and annotated their *mathematical graphics* we read more research and theory and, keeping emergent writing in our minds, began to question whether it would be possible to trace the trajectory of young children's developing understanding of the symbolic written language of mathematics through their *mathematical graphics*.

Research by Hughes, published in *Children and Number: Difficulties in Learning Mathematics* (1986) focused largely on children's representations of small quantities and a small number of examples of addition and subtraction: we now had numerous examples of many other aspects of number and calculations, and of all other areas of mathematics. We wondered how we might evaluate the numerous examples we had amassed and if it would be possible to trace some development of understanding through them. Since there was nothing published that we could turn to, we began to consider how we might analyse them. After a great deal of discussion over time we developed our taxonomy, showing how children's own symbols and representations support deep levels of mathematical thinking (Carruthers and Worthington 2005a, 2006) and this stimulated increasing interest from early childhood teachers and educators, and subsequently influenced government policy (e.g. DCSF 2008a, 2008b). While we had examples of all areas of mathematics, since number is integral to them all the taxonomy itself focuses on number and quantities and children's own methods for calculations (see page 76).

Meanwhile, we were also both working on research for our masters degrees, one a parent–child study of a young child's developing mathematical understanding between the ages of 22 and 42 months, and the other analysing cognitive challenge in child-initiated play. During the 1990s we conducted a number of studies on children's mathematics, ranging from short to year-long studies. With the children we taught (and their families) we researched mathematical schemas. In another study we explored parents' own feelings about mathematics and their memories of learning the subject, linking these with the maths in which they engaged at home as adults, and the maths they saw their own children explore at home within their play (see Carruthers and Worthington 2006).

We conducted two studies with teachers and practitioners working with children from 3 to 8 years throughout England (each with approximately 300 individuals), the first exploring their practice in relation to teaching 'written' maths and the second focusing on teachers' and practitioners' understandings of 'creativity' in mathematics. Other studies included adult modelling of written mathematics and its impact on children's symbolic tool use in mathematics (for details of these studies see Carruthers and Worthington 2006 and the Children's Mathematics Network at www.childrens-mathematics.net.

Emergent mathematics?

Children's mathematical graphics have been compared to 'emergent writing', although they are not the same: the subject content, the children's thinking about the symbolic written languages of 'writing' and 'mathematics' and the ways in which they represent their ideas are inherently different. It is also important to avoid any suggestion that children's early written notation 'just emerges'. However, both emergent writing and *children's mathematical graphics* share one significant aspect, in that children make and attach meanings to the graphical marks and the symbols they choose to use. We originated the term *'children's mathematical graphics'* (now used in official guidance in England for teachers and practitioners, e.g. DCSF 2008a, 2008b, 2009b), and strongly believe that it is important to include the word 'children's' since (unlike copying, colouring-in or completing a worksheet) the graphics, the mathematical thinking, the representations and the meanings are the children's own.

Our current research reveals that above all, it is the cultural conditions that teachers and practitioners create that nurture the extent to which children will choose explore and represent personal mathematical meanings. The learning culture also impacts on the extent to which children's play and graphics are valued and understood (Worthington 2010a).

Written mathematics in early childhood education in England

It is difficult to imagine that writing would be omitted from communication language and literacy in the EYFS, or from the literacy section of the Primary Strategy, yet effectively the beginnings and early development of 'written' mathematics have been omitted from the mathematics areas of both curricula. In spite of some positive statements in both the National Numeracy Strategy's booklet *Teaching Written Calculations* (QCA 1999), and in the *Curriculum Guidance for the Foundation Stage* (QCA 2000), both lack examples and guidance on this important aspect of mathematics.

With the establishment of the EYFS the guidance recommended: 'Value children's own graphic and practical explorations of *Problem Solving, Reasoning and Numeracy*' (DfES 2007a: 61), advising that children 'create and experiment with symbols and marks' and 'begin to represent numbers, using fingers, marks on paper or pictures' (2007a: 64–5). However, the guidance fails to show any relationship between what is essentially 'handwriting' (written number formation), 'marks on paper and pictures' and children's understanding of the symbolic written language of mathematics. Selter (1998: 1) observes that 'there rarely seems to be any disagreement about the fact that teaching should build on children's mathematics, however, there are different opinions about how this could be realized in the classroom'.

In England the Department for Children, Schools and Families (DCSF) commissioned a two-year review of mathematics teaching in early years settings and primary schools, and in 2008 the final report of the 'Williams Maths Review' was published (DCSF 2008a). The report's authors emphasized that 'The review also lays great store by play-based learning of a mathematical nature, and makes specific recommendations regarding early mark-making as a precursor to abstract mathematical symbolism' (page 4). It includes a dedicated section on *children's mathematical graphics* and proposes a number of ways in which those working with young children can 'implement effective early years pedagogy' including 'A culture with a significant focus on mathematical mark-making' and 'a learning environment that encourages children to choose to use their own *mathematical graphics* to support their mathematical thinking and processes' (page 37). This publication led to two further booklets, *Mark Making Matters* (DCSF 2008b) and *Children Thinking Mathematically: PSRN Essential Knowledge for Early Years Practitioners* (DCSF 2009b), which we were commissioned to write.

'Recording maths', or representing mathematical thinking?

In England children's written mathematics is commonly referred to as 'recording': an activity that describes children recording something they have already done in a practical context, or filling in answers that they have already worked out mentally. In contrast, the emphasis in *children's mathematical graphics* is on the *processes* of mathematics, such as negotiating and co-constructing meanings, reasoning, generalizing and on children using their own methods to solve problems through their personal, graphical representations (QCA 2003; DfES 2006). Building on their earliest symbolic representations in their play and graphicacy helps children to 'translate' between their early informal marks and the more 'standard' symbols and written language of school mathematics, supporting their understanding at a deep level.

In a sense, *children's mathematical graphics* are their mental methods on paper: children need to be free to choose how they will represent their mathematical thinking in a way that best fits their purpose, the particular mathematical context or the calculation they are exploring. Real contexts will allow children to make greater sense of their mathematics when they explore their thinking through their own representations.

Referring to Seeger (1998), Pape and Tchoshanov argue that '. . . representations must be viewed as vehicles for exploration within social contexts that allow for multiple understandings of mathematical content. This conceptualization necessitates an alternative view of the use of representation from that which is 'typical within mathematics classrooms' (2001: 124). DiSessa *et al.* (1991: 156) argue that students' own representation has a number of potential advantages:

> first, it closes the gap between prior knowledge and the material they are involved with. Second, it provides opportunities for creative engagement and ownership of conceptually difficult material. And third, it enables students to exercise their meta-representational knowledge, which is expected to be of value in the creation of new representations.

Recording what they did following a practical activity has limited value for children and involves lower levels of thinking. Children do not need to record mathematics if they can do it mentally; neither do they need to record something they have worked out in a practical context. Recording places the emphasis on marks and drawings as a *product* and has a lower level of cognitive demand (thinking) in mathematics. The difference between *representing mathematical thinking* and *recording maths* is one of quality and depth of thinking. However, diSessa notes 'how rare it is to find instruction that trusts students to create their own representations' (1991: 156). This view is further reflected in a recent paper by Terwel *et al.* who propose that 'Although there have been positive changes in the past decades, we believe that today, diSessa's statement holds true for many classroom practices' (2009: 28–9).

The taxonomy and dimensions of children's mathematical graphics

The taxonomy (see Figure 6.8) provides a system that classifies *children's mathematical graphics* according to their functions and the child's intended purposes. It is generally possible to match a single example to one of the dimensions (either *written number and quantities*, or *children's own methods for calculations*) and then to see what other dimensions may be linked. Throughout their *mathematical graphics* children are also continually *exploring symbols*.

The taxonomy is non-hierarchical and the various dimensions[1] should not be seen as 'stages' through which all children pass. Its value is in helping teachers and practitioners to understand *children's mathematical graphics* and teachers should not teach to the taxonomy. They should however find it invaluable in supporting and informing their assessment.

Written number and quantities

Early exploration with marks: attaching mathematical meanings
Children use their own early marks and attach mathematical meanings to them.

Explorations with symbols
Children use a range of graphical symbols to support and communicate their mathematical thinking.

---

[1] The dimensions of the taxonomy were first introduced in Worthington, M. and Carruthers, E. (2003) *Children's Mathematics: Making Marks, Making Meaning*. London: Paul Chapman.

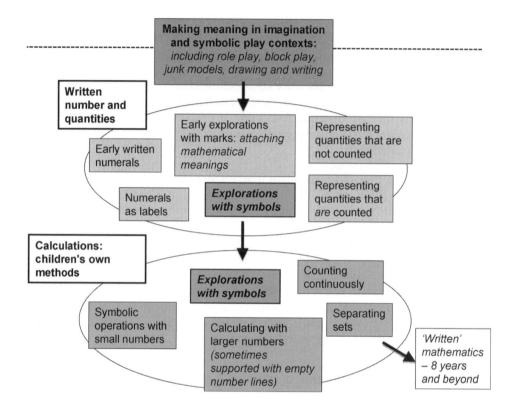

Figure 6.8 Taxonomy: charting children's mathematical graphics (birth to 8 years)

Early written numerals
Children refer to their marks and symbols as numbers as they explore ways of writing them.

Numerals as labels
This is when children use numbers they see in their environment as labels to signify either a quantity or number. Knowing and talking about numbers is different from writing them and when they choose to write numbers in context, children have converted what they have read and understood them as standard mathematical signs.

Representing quantities that are *not* counted
Children either attend to the link between their early marks and meaning as a quantity in a general sense, or use the paper to arrange items and numerals in a general sense (but without counting them).

Representing quantities that *are* counted

During the same period that children represent uncounted quantities, they may also represent quantities that are counted: this leads to early calculations and all other aspects of written mathematics.

Calculations: children's own methods

Counting continuously

This is when children represent two sets of objects, icons or numerals they are adding in a horizontal (or occasionally vertical) arrangement on the paper, in a continuous line. The two sets are not separated and they count items continuously to arrive at a total (rather than add). They understand the need to count everything to arrive at a total.

Separating sets

Children use a range of strategies to show that the two amounts they are adding (or subtracting) are separate – for example, perhaps by leaving a space between them, by separating the sets with words or putting a vertical line between them.

Explorations with symbols

Children explore the role and use of symbols by using personal or invented symbols, or approximations of the standard symbols.

Symbolic operations with small numbers

Children's progression in written mathematics, from their earliest marks to which they attach mathematical meanings to calculations with standard symbols, is a continuum: their choice to represent calculations in their own ways shows how they understand what they do.

Calculating with larger numbers

Calculating with larger numbers is more challenging, since children may need to manipulate several steps and mental methods, and the taught 'empty number line' can be of value. Strategies children use include using known number facts, counting on from the larger of the two numbers and partitioning numbers. Children need extended periods of time in which to explore symbols in their own ways, before they are ready to use 'standard' symbolic operations with small numbers. By around the age 7 most children will be using standard mathematical symbols and horizontal written calculations by choice.

Additional strategies

Implicit symbols

Children may sometimes show they have some understanding of signs such as '+' and '–', although they have not represented the symbols: leaving a space between the two sets to be added (or before the total) shows that the symbol is *implied* and that they understand the calculation. They may 'read' their calculation out loud as though to include written features which are absent (for examples see Figures 2.11 and 2.12).

Code switching

Children often combine personal and standard ways of representing calculations as they think about their representations and gradually integrate standard symbols as they move towards increasingly efficient methods.

Narrative action

Narrative action is a particularly significant feature and describes individuals' decisions to include either hands or arrows in their calculation to signify the operation of either addition or subtraction (see Figures 2.15 and 2.16). When children use narrative action in this way, they appear to have developed an internalized *mental image* of earlier concrete operations of addition or calculation. In these instances the hands or arrows allow them to reflect on the operant and its function. Narrative actions have an important 'bridging function' between concrete and mental operations, between physical actions with concrete materials and the operators within written calculations, enabling children to understand what their signs signify and their active function within a calculation.

Conclusion

Teachers' skilled observations and their ability to reflect on them, coupled with their ability to scaffold children's understandings and their engagement in collaborative discussions to negotiate and co-construct meanings, support *children's mathematical graphics*. Above all it is the positive learning culture that shows the extent to which adults truly value children's meaning-making and that makes the greatest difference to children's understanding.

The next three chapters focus specifically on *exploring and communicating mathematical thinking* through graphical representations in a range of contexts, showing young children's incredible ability to use graphical symbols to make personal mathematical meanings. Chapter 7 includes a range of observations of children using the*ir mathematical graphics i*n play. It includes some brief, spontaneous moments of mathematical enquiry, and sustained periods of play where children explore and develop their interests and mathematical thinking in depth.

Reflections

- Observe children playing and note any who use mathematical graphics within their play. Talk to individuals, valuing their representations and noting anything they say about their marks and symbols.
- Collect several examples of children's mathematical graphics and together with your colleagues use the taxonomy to discuss what these examples tell you.
- Begin a portfolio of children's mathematical graphics to help you develop your understanding of how graphics are used to support and communicate children's mathematical thinking.

Recommended reading

Carruthers, E. and Worthington, M. (2006) *Children's Mathematics: Making Marks, Making Meaning*, 2nd edn. London: Sage.

DCSF (Department for Children, Schools and Families) (2009) *Children Thinking Mathematically: PSRN Essential Knowledge for Early Years Practitioners*. London: DCSF.

7

Children's mathematical graphics in play

'I'm tall Maria. Look how tall I am. I'm getting taller. We're the same tall.' 'Uh-uh Mollie. I'm five and you're three.' I'm taller Maria. Every day I eat. . . . I'm this tall too. I'm five inches.'

(Paley 1986a: 118)

What is this chapter about?

- **Case Studies** Children exploring their thinking through their mathematical graphics in play, illustrated through both short and long case studies.
- **Sustained periods of play** and mathematical graphics.

This chapter includes a range of case studies showing children aged 3–5 using mathematical graphics in self-initiated play. The case studies show how mathematics is often embedded in, and emerges through, play – not as an adult-planned learning 'objective' for mathematics, but because it is integral to the children's play context and cultural interests, and has personal meaning for them. The studies demonstrate children playing with mathematical ideas and exploring and communicating their mathematical thinking through their graphical representations.

In England, a recent report by the Qualifications and Curriculum Development Agency (QCDA) surveyed 1,200 early years teachers and practitioners, and found that one of the specific areas causing difficulties was mathematics (referred to in the current curriculum as 'problem solving, reasoning and numeracy' or PSRN) (DfES 2007a). The most challenging aspect of maths teaching cited was 'to deliver aspects of PSRN through child-initiated activities' (QDCA 2010: 8). Unfortunately the report failed to use the word 'play' anywhere.

Case studies

In all the play episodes that follow, the examples of *children's mathematical graphics* are explained with reference to the taxonomy in Chapter 6 (see Figure 6.8). The first group (Figures 7.1–7.6) captures spontaneous moments of mathematical enquiry, showing how the children draw on and develop their knowledge of quantities, counting and written numerals. While they are brief, such mathematical episodes are worth capturing as they provide teachers with valuable knowledge about aspects of children's mathematical understanding as they 'use and apply' it within their play.

CASE STUDY

Chevaun's 'oven numbers'

Child

Chevaun (2 years, 11 months)

Teacher

Hugo

Context

Playing outside

The mathematics

Numbers

Dimension of the taxonomy

Written number and quantities: *early exploration with marks – attaching mathematical meanings*

Chevaun was playing outside at the kitchen and read her marks to her key worker: 'I'm doing labels. I'm doing "4" and "18" on the oven.' The result is shown in Figure 7.1.

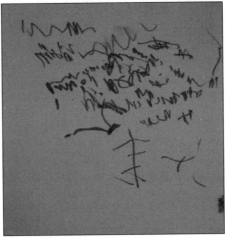

Figure 7.1 Chevaun's 'oven numbers'

CASE STUDY

'My dad's got five eyes!'

Child

Shakkai (4 years, 10 months)

Teacher

Kylie

Context

Shakkai and his friend were drawing their dads

The mathematics

Counting, addition

Dimensions of the taxonomy

Written number and quantities: *representing quantities that are counted.*
Calculations – *children's own methods: counting continuously*

Shakkai drew four eyes on his dad's face and the boys laughed at their drawings. Shakkai said, 'I am going to add another eye. Look! He has five now!' (see Figure 7.2).

Figure 7.2 Shakkai's dad

'Four for Kieran'

Child

Sophia (3 years, 9 months)

Teacher

Sue

Context

Talk time

The mathematics

Written numerals

Dimension of the taxonomy

Written number and quantities: *early written numerals*

At talk time Sophia decided to write several numeral fours, repeating '4, 4, 4' and explaining 'four for Kieran' (see Figure 7.3). She was not yet 4 years old and seemed intrigued with the idea of attaining this age.

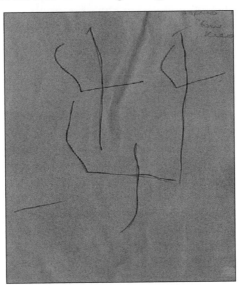

Figure 7.3 'Four for Kieran'

CASE STUDY

'Chocolate biscuits, rice pops, sausages'

Child

Amelia (3 years, 4 months)

Teachers

Donna, Jo and Keeley

Context

Playing in the shop

The mathematics

Counting

Dimension of the taxonomy

Written number and quantities: *exploring symbols; representing quantities that are not counted*

Amelia was busy organizing things, chatting away happily to her friends. She appeared to be thinking about things she liked to eat and items that her mum bought when they went to the supermarket together (see Figure 7.4).

Figure 7.4 'Chocolate biscuits, rice pops, sausages'

Harry's phone number

Child

Harry (4 years, 7 months)

Teacher

Emma B.

Context

Harry in the painting area

The mathematics

Numbers

Dimension of the taxonomy

Written number and quantities: *numerals as labels*

Harry pointed to the paper (see Figure 7.5) and said, 'This is where you talk.' He then painted the numbers '1', '2', '3', explaining, 'This is a telephone. I'm going to phone my mum; her number is 123321.'

Figure 7.5 Harry's phone number

CASE STUDY

Shazil's climbing frame

Child

Shazil (4 years, 7 months)

Teacher

Stephanie

Context

Shazil had just been on the climbing frame

The mathematics

Space and shape, number

Dimension of the taxonomy

Written number and quantities: numerals as labels

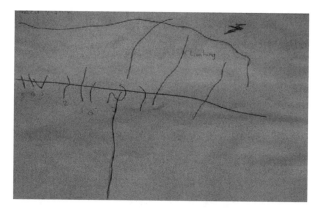

Figure 7.6 Shazil's climbing frame

Shazil drew the graphic shown in Figure 7.6 outside on the picnic bench. As he made the small vertical lines he said, 'Five, four, three, two, one, six and seven.'

The next set of case studies focus on some longer episodes of child-initiated play, showing how the children explored mathematics and mathematical symbols.

Stacey's café

Child

Stacey (3 years, 2 months)

Teacher

Emma B.

Context

Outside play house

The mathematics

Money

Dimension of the taxonomy

Written number and quantities: *early exploration with marks – attaching mathematical meanings*

Stacy was playing in the little house in the outside play area. Picking up a pen and a notebook she asked her friend, 'What do you like?' Her friend replied, 'Chicken tikka please.' Stacey asked, 'Chicken and chips?' and filled the page in the notebook with large circular shapes, small dots and lines, and then shaded part of the central area (see Figure 7.7). Stacey asked her friend for 'Fifty pounds please' and her friend pretended to give her some money, which Stacey put in her pocket. Stacey drew on her home knowledge of ordering a take-away meal, clear in her understanding that orders were written down in the restaurant and integrating her understanding of asking for and taking orders.

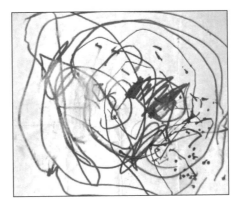

Figure 7.7 Stacey's order

CASE STUDY

'How old are you?'

Child

Baylee (5 years, 0 months)

Teacher

Kylie

Context

Baylee had just had her birthday and wanted to find out how old the other children were

The mathematics

Counting, data collection

Dimension of the taxonomy

Written number and quantities: *numerals as labels, representing quantities that are counted*

Baylee wrote the numbers 1–10 and after she had begun writing ticks, decided to put a dot beneath each (as if to make it easier to see where the tick went) (see Figure 7.8). Beginning with ticks beneath '5' and '7' she then went around the class with her clipboard, asking how old the children were. When she came to the teacher she had ticks beneath '7' and '2' and explained her brother was '7' and her friend's sister was '2'. Baylee asked her teacher how old *she* was, and thinking hard about the number, wrote it down and then gave it a tick. Baylee clearly thought about layout and ordered the data as she collected it, making her findings easier to read and discuss (or analyse) afterwards.

The next day Sian (who had been watching Baylee) decided to use her idea: she wrote the numbers 1–54 (see Figure 7.9) and asked her teacher how old she was before asking the children their ages. Drawing on Baylee's use of ticks, she was able to explain her findings and said that there were five children aged 4 years.

Age is of great interest to young children and, as this example shows, having clipboards available encourages them to spontaneously collect data about things of personal interest. Sian's data (drawing on what Baylee had done) is also a good example of *peer modelling* (see Chapter 11).

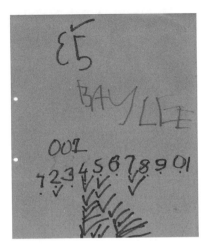

Figure 7.8 Baylee: 'How old are you?'

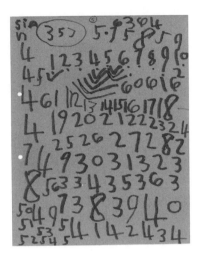

Figure 7.9 Sian asks the children their ages

CASE STUDY

61,000,000 – 'This one's the biggest!'

Children

Francine (4 years, 5 months) and Hayden (4 years, 8 months)

Teacher

Maxine

Context

The children chose to play with some old diaries and notebooks at the graphics table, making various marks and writing numbers

The mathematics

Writing numbers, counting

Dimension of the taxonomy

Written number and quantities: *numerals as labels, representing quantities that are counted*

Francine was very excited that she could write a million and, as she wrote each line of numerals, read them in a loud voice (see Figure 7.10). Several other children joined her, keen to write equally large numbers and Francine reminded them that they needed 'six zeros'. Francine explained to her teacher that 'This

Figure 7.10 'Millions and millions!'

Figure 7.11 Hayden's numbers

one,' pointing to '61000000', 'is the biggest!' and other children joined in to discuss who had written the largest number. While this does not imply that Francine had a full understanding of our number system into the millions, there is every reason to support young children's curiosity and interest in numbers that extend beyond 10 or 100, and how they are written. Meanwhile Hayden appeared to be to writing a range of numbers he knew (see Figure 7.11). As he wrote he occasionally pointed to a number, explaining that it was his brother's age, or the age of his friend.

CASE STUDY

'It tells you how much medicine you're giving'
Children

Kiera (5 years, 2 months) and Lauren (5 years, 0 months)

Teacher

Karen

Context

Imaginative play – 'doctors'

The mathematics

Numbers, quantity, measurement

Dimension of the taxonomy

Written number and quantities: *numerals as labels*

Karen explained: 'Kiera was enjoying being a doctor, helping Tanny and Lauren' (see Figure 7.12). 'She gave some medicine to Tanny, explaining that the number on the cup said "80". "It tells you how much medicine you are giving." Lauren also enjoyed being a mummy, using the phone, making a telephone number list. She wrote and read "300" for the doctor and "37", this time saying "three" and "seven" for Oma.'

Mathematics is often naturally embedded in children's self-initiated role play situations as they explore and draw on personal experiences as home. Their pretend play provides meaningful contexts for mathematics such as measuring liquid medicine and telephone numbers.

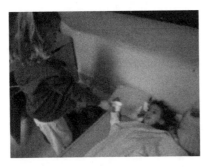

Figure 7.12 Giving medicine to the baby

Figure 7.13 Lauren's list of phone numbers

CASE STUDY

'Hello Kitty!'

Child

Gaya (4 years, 6 months)

Teacher

Emma H.

Context

Gaya was thinking about her granddad's forthcoming birthday

The mathematics

Numbers (age-related), quantities

Dimension of the taxonomy

Written number and quantities: *representing quantities that are not counted*

Drawing around a small cutting board of the shape she wanted, Gaya had drawn a 'Hello Kitty' cake to represent a birthday cake for her granddad. She explained, 'Granpa's really old – he's 30,' and while Gaya knew her granddad's age, she drew just a small number of 'candles' (the small circular marks) as a generalized quantity (see Figure 7.14). Later she explained to her teacher, Emma, that 'He *really* has to have 30 candles.' Drawing on her own experiences of birthdays and celebrations with birthday cakes, she understood that the quantity of candles on a birthday cake should match the person's age. Gaya's interest in her granddad's age and in birthdays were linked with her love of 'Hello Kitty', a character from popular culture that originates in Japan. It seemed logical to her that since she loves this character her granddad would as well, and that this would be the birthday cake he would most want. Children benefit from choosing how they will represent their mathematical thinking so that it best fits the meanings they want to communicate. While other children might have drawn 30 candles, or written words or numerals, at that moment Gaya's generalized representation of 'lots of candles' suited her purpose at the time.

Figure 7.14 Gaya's 'Hello Kitty' cake

CASE STUDY

Rory's day

Child

Rory (4 years, 9 months)

Teacher

Emma H.

Context

Rory's dad had been away with work for several weeks and Rory was thinking of what he would be doing prior to his dad's arrival the next day.

The mathematics

Measuring time, sequencing and predicting events

Dimension of the taxonomy

Written number and quantities: *representing quantities that are not counted, representing quantities that are counted*

In the centre-left of Figure 7.15 the two circles represent the snack and lunch he would eat. Two irregular shapes with vertical lines represent trees outside, by the school fence, and the playground where he would be playing. Beneath them he drew the bus, explaining the time he left to come to school and that there would be lots of people on the bus. On the right of the page the circles and lines represent the plane on which his dad would be returning home. Rory included events that held particular significance for him, enabling him to anticipate and explore the events of the following day. His drawings and symbols have a temporal quality, allowing him to order events that would precede his dad's arrival by plane when they would be reunited. Although Rory's representation arose through a personal, emotional need, it also enabled him to explore aspects of mathematics.

Figure 7.15 Rory's day

CASE STUDY

'Do not move my logs'

Child

Tim (4 years, 1 month)

Teachers

Ros and Michelle

Context

Playing outside

The mathematics

Counting, numerals and length

Dimension of the taxonomy

Written number and quantities: *early written numerals, representing quantities that are counted*

Tim was in the garden, lifting some old fence posts onto a trolley. He spent a lot of time transporting them to another area of the garden and his teacher joined him, wondering how many logs he had moved. Tim was too engrossed in what he was doing to reply, but later, taking one of the clipboards and a pen from the basket outside, he drew circular marks and other symbols in a linear arrangement across the top of his page (see Figure 7.16). His teacher watched as he counted the marks (from left to right) and his teacher wrote the numbers he said: when he reached 'eight', he returned to the beginning to check, counting each in turn as he touched them, and then continuing to 25 (he missed a few). Next Tim drew four long logs, saying, "They're very lo-o-o-ong.' Beneath the first 'long log' he wrote a sign and read, 'This says "do not move my logs and do not park here".' Tim's experience of physically lifting and moving his logs was clearly important to him and his decision to represent and communicate this thinking revealed his mathematical understandings.

Figure 7.16 Tim's logs

Mimi's crosses

Child

Mimi (3 years, 4 months)

Teacher

Hugo

Context

Playing outside

The mathematics

Quantities, counting, shape and space

Dimension of the taxonomy

Written number and quantities: *exploring symbols, representing quantities that are not counted*

Mimi made a series of crosses in chalk on the ground and surrounded each one (with the exception of two) with a border (see Figure 7.17). 'Now everyone can no come,' she said. Elizabeth approached Mimi, walking across her marks. 'No! She's come.' Mimi drew a boundary around the two crosses to emphasize the point that she wanted the space kept free of others (see Figure 7.18). Elizabeth soon returned and bounced a ball on each cross. Mimi objected, saying 'Hey, don't go inside!'

Figure 7.17 Mimi's crosses   **Figure 7.18** Enclosed crosses

After an intervention from an adult, Elizabeth took her own chalk and drew a long line that passed by Mimi's marks. In response, Mimi encircled all her own marks, creating a wide boundary (see Figure 7.19). 'I did a long line now,' she explained as she stood in the large space, as if on guard (see Figure 7.20).

Figure 7.19 Mimi separates her crosses from Elizabeth's

Figure 7.20 'I did a long line now'

Mimi used graphical symbols and boundaries to get her message across to others. She chose an 'x' to denote a denial of access and reinforced this by enclosing each 'x' individually, with four of these enclosed crosses forming a square. She then enclosed two crosses together with a single border.

Finally, Mimi enclosed the entire space to create a substantially large area. She added various additional shapes and named them ('a rectangle', 'a diamond'), linking this to her experience by referring to the computer she had at home. She then represented a rectangular outline connected with a single line to an irregular enclosure, clearly showing a computer with a keyboard attached (see Figure 7.21).

Figure 7.21 Mimi's computer

Sustained periods of play

Here we focus on periods of sustained play where children explored and developed their interests and mathematical thinking in depth. The case studies show the collaborative, social nature of play that allows children to co-construct their understandings and develop deep levels of understanding as they draw on their home knowledge (their 'funds of knowledge'), and serve to highlight the adults' role in supporting their thinking. Some of the observations show how the children's interests developed over extended periods of time.

CASE STUDY

The children's television

Children

Jaydon (3 years, 3 months), Hazel (4 years, 6 months), Abdullahi (4 years, 4 months), Nathan (4 years, 7 months) and other children in the nursery who joined in for short periods

Teacher

Emma B.

Context

Spontaneous play outside and in. This was triggered by a piece of rectangular card that had a large rectangle cut from it: the children thought it looked like a television and this sparked a whole episode of continuous play for several weeks

The mathematics

Numerals as labels

Dimension of the taxonomy

Written number and quantities: *early written numerals, numerals as labels*

Jaydon looked at their pretend television (see Figure 7.22), commenting, 'There is no plug in.' He then got a block to represent a plug, explaining, 'It's sorted now.' He drew the lines on the children's television to create wires on the front and the back of the set. Jaydon is very interested in how things work and is aware of the wiring and electrical connections of televisions.

Figure 7.22 Nathan and Abdullahi play with the children's television

Figure 7.23 Jaydon using a real spirit level to test if the television is straight

Figure 7.24 Abdullahi writing numerals for his remote control

The children decided to put their television in the gazebo but it kept blowing over. Abdullahi solved this problem by taping it down. Jaydon noticed that it wasn't straight so he got a real spirit level to check (Figure 7.23), saying, 'No, it's not straight, gotta straighten it up.'

Hazel became very interested in the play episodes, concentrating on the journeys of going out (of the gazebo) and coming 'home', relaxing in front of the television with family and friends.

Abdullahi created a remote control for the television (Figure 7.24) using a piece of paper and reading his marks as numerals (buttons) to pretend to change the channels of his television. He said, 'I want to watch ballet, it's one, four, zero. Five, one, zero is our favourite, five, four, two is good home cartoons.' Later he folded his paper remote control into a different shape to transform it into a PlayStation controller. He said, 'We got a PlayStation at home, a car one. I winned the race when I played with Sara and Samira.'

Hazel and the others also became interested in the remote control idea. Hazel said, 'I am using a pen as a controller. I want to watch *Boomerang* on zero, seven, zero, seven, zero. *Bat Girl* is zero, seven, three, zero, seven, three.'

Two weeks after making his first remote creation, Abdullahi was still engaged with his television play. He decided to make three remotes, all with wavy-style marks, cut them out and gave them to his friends.

This very rich and sustained play period captured the children's imagination and drew on their knowledge of technologies which they threaded through their play over several weeks. Emma, the children's teacher and key person who captured these young children's play experiences, commented that this episode 'Opened up a new pathway of interaction based on shared experience of television at home; valuing the experiences children bring to nursery.'

The children were weaving in and out of different modes of representation and thinking deeply: it is clear that children draw on their considerable knowledge of new technologies and media, and that '. . . the potentials of these [new] technologies imply a radical social change, a redistribution of semiotic power' (Kress

2003: 17). Hazel preferred to use a pen for her remote control but Abdullahi used paper and pencil. Children will use whatever medium suitable for their purposes at the time (Kress 1997).

The children's graphics were an integral part of their play – for example, the drawings of the wires and the numerals written on the paper to represent buttons on the remote control. Their graphics did not stand alone but were a vehicle that supported their play. The culture within the nursery values and gives opportunities for children to use their graphics; it is therefore natural for children to pick up graphical tools in their play and, as they increasingly realize that their marks and symbols support their thinking, the more they use them.

CASE STUDY

Paper calculators[1]

Children

Mason (4 years, 5 months) and Alfie (4 years, 3 months)

Teacher

Carole

Context

Quiet time for all-day children in the nursery

The mathematics

Counting

Dimension of the taxonomy

Written number and quantities: *early written numerals; numerals as labels*

This play began several weeks earlier when Mason played with a real calculator: he seemed to be using it as a digital game, pressing the buttons and commenting excitedly, 'Fighting games! Video games!' It was Mason who first decided to use a small notebook to make 'paper calculators', explaining as he tore off a sheet, 'This is a *different* calculator with computer games on' (see Figure 7.25). Alfie watched Mason use pages from a notebook to make 'calculators' and decided to join in. Having drawn shapes on his page, Alfie ripped it off the notepad and then made more symbols, saying, 'Six, seven, eight, nine, I've done a number ten,' followed by a third sheet, announcing, 'Nine, ten, eleven, twelve.' In 20 minutes Alfie made a total of seven paper 'calculators' (see Figure 7.26).

[1] This case study was first published in Worthington, M. (2010) Play is a complex landscape: imagination and symbolic meanings, in P. Broadhead, J. Howard and E. Wood (eds) *Play and Learning in the Early Years*. London: Sage Publications, and is included here with kind permission of the publishers.

Figure 7.25 Mason making a paper calculator

Figure 7.26 One of Alfie's paper calculators

Figure 7.27 Mason: 'This is a *different* calculator, with Batman on'

Several of Mason's peers joined in and developed their own 'paper calculators', and making another, Alfie announced, 'Lots of fighting!' A week later Mason explained he'd made a calculator with 'Batman' on it (see Figure 7.27).

The boys returned again and again to this play over a period of two terms, making rapid scribble marks, drawing 'buttons' on their 'calculators' (and pressing them), and writing numerals. This example shows the depth of children's thinking when they are free to follow their own interests, and when their voices are heard and valued. Their play allowed the children to adapt, co-construct and negotiate symbolic meanings so that layers of meanings evolved over time.

'We're building a house!'

Children

Jaydon (3 years, 6 months) and Isaac (3 years, 7 months)

Teacher

Emma B.

Context

Free play in the block area in a nursery setting. For several months these children were interested in the blocks, creating and re-creating imaginary worlds. They often spent days on the same theme, bringing ideas from home and extending their thinking on the subject of their play

The mathematics

Mapping, directions, space, shape and size

Dimension of the taxonomy

Written number and quantities: *explorations with symbols*

Jaydon selected some large paper and a pen in the block area and began making very large marks: 'We're building a house; the big huge television is going in this room.' Isaac joined Jaydon and also added his own graphics on the paper: 'This is my big eating room; here's a nice clean room and kitchen.' Jaydon exclaimed, 'My house is getting bigger and bigger . . . This is the roof.' Jaydon and Isaac agreed it would be useful to display their building plan on the wall so they could refer to it when building (see Figure 7.28). Asking for more paper, Jaydon was motivated to make intersecting lines, saying, 'This is to stop people going past; cones will go here, not a "no entry" sign.'

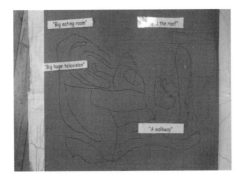

Figure 7.28 The house plan

Using the blocks, Jaydon decided to create a roof by making an enclosure, and systematically fitted blocks inside. He was very clear about the way in which the blocks should be placed to maximize the use of space. Later, Isaac (who is interested in electricity) said, 'These are electric steps; they generate electricity and could give you an electric shock!' Isaac and Jaydon played together to make their kitchen. Using one of the blocks as a sort of nail gun, Isaac said, 'This shoots out nails, to fit the shelves, nail, nail, nail.' Jaydon joined him, using a plank to represent a drill: 'Brrrrrrrr. It's all fitting together.'

When analysing the children's graphics, Emma commented: 'These spontaneous drawings offer a different medium for representing and Jaydon and Isaac seemed to understand the use of making marks prior to this self-initiated play project. They were intrinsically motivated to make marks collaboratively, narrating their marks as they did them. Their marks were mathematical in the way they represented a plan or map of a house, incorporating size: as their house and rooms became bigger so did their marks. When talking to Isaac's mum and dad they said that as his dad was a carpenter Isaac had watched him fit a kitchen. Isaac had obviously used this knowledge in his construction.'

The boys' use of the blocks to represent not only their building but also the tools to build, showed that they were working on many abstract levels, firstly using their graphics and thinking to plan their construction and then using the blocks to represent all their ideas. Jaydon's use of the plank for the drill shows that he understood he could adapt something available to *represent* his ideas (a plank does not look like a drill). He was able to think beyond the literal: he just needed an object that would *signify* a drill and the details were in his mind (see Chapter 2).

Blocks have long been one of the bedrocks of essential equipment in children's play (see e.g. Gura 1992) since children can use them in open ways to signify anything they want because of their simple shapes and sizes. Blocks are useful to children in a variety of contexts, not only for building structures but also in their imaginary play: they do not suggest any particular object and may make it easier for children to apply their own ideas. It is important that children have a range of equipment to use and combine in their own ways to make and represent meanings, for this freedom helps them think and formulate ideas and wonderful 'playscapes'.

CASE STUDY

'Look! No chicken!'

Child

Shereen (3 years, 5 months)

Teacher

Emma B.

Context

Nursery; children's kitchen area

The mathematics

Money

Dimension of the taxonomy

Written number and quantities: *exploring symbols; representing quantities that are not counted*

Taking a lump of playdough, Shereen turned to one of her peers saying, 'We make cake, you want cake?', and fetching four bowls she set them out in front of her. She divided the playdough into the four bowls, putting smaller amounts in the two smaller bowls and then putting her 'cakes' in the toy oven (a cardboard box). She returned to her table, setting out more bowls and cutlery. Returning to the oven, she took out the bowls asking, 'Who's eating?' She gave a bowl to each child sitting around the table and said, 'Eating time.' She allowed the children time to pretend to eat before collecting their bowls: one child had not joined in with pretending to eat and as Shereen picked up her bowl she looked into it asking, 'You not very hungry?'

Picking up the pad of paper and pen from the table, Shereen began drawing wavy lines on the lined paper, saying to her teacher, 'I'm writing chocolate bar, what you want? I've got rice, chocolate, chicken?' (see Figure 7.29). Emma replied that she'd have some rice and Shereen said 'Okay', busying herself with the play-dough and pretending to prepare the rice. 'There you go, it's two, one, two,' she said, referring to the price of the rice (see Figure 7.30). Emma pretended to eat the rice and then Shereen returned with her notepad to see if she wanted anything else to eat. 'What you want: rice, chocolate, cake and chicken?' Emma said that

Figure 7.29 'I'm writing chocolate bar, what you want?'

Figure 7.30 'Okay. There you go, it's two, one, two'

she didn't want chicken and Shereen wrote something for 'chicken' and put a cross by it: 'It says "x" – no chicken.'

After a while Shereen returned once again to Emma, to see if she would like to order more food and this time Emma said that she would have some chicken. Looking at her notepad Shereen pointed to the 'x' she had written: 'Look! No chicken! You want mushroom?' Emma agreed that she'd have mushroom instead. Shereen again wrote a cross by her marks for chicken, explaining, 'Look. A tick, that mean we got some,' then added, 'you want ice cream? It's three, four.'

The next day Shereen continued her café play outside, once again taking orders for food (Figures 7.31, 7.32).

This rich observation reveals a great deal about Shereen's symbolic knowledge as she drew on her home experiences where serving and sharing meals is culturally significant, including eating out with her family. Deciding to take orders from her 'customers' enabled Shereen to explore the role and function of abstract symbols and use them to make and communicate personal meanings.

Figure 7.31 Shereen taking orders

Figure 7.32 Cooking the rice

CASE STUDY

Car park at the beach

Children

Daniel (3 years, 4 months) and Isaac (3 years, 6 months)

Teacher

Emma B.

Context

Block play

The mathematics

Mapping, direction, space, size, money, time and numbers

Dimension of the taxonomy

Written number and quantities: *early explorations with marks – attaching mathematical meanings; exploring symbols*

Isaac selected a large piece of paper and together he and Daniel created a road scene. Their marks represent a beach, a car park with spaces, fences, crossroads, a helipad, mud and a gate that can open and close. Isaac explained: 'This is the car park gate, the square on the outside means it's shut.' The following day he 'opened' the gate by making the two horizontal lines.

After playing with the small toy car and lorries, Isaac decided to create a road map for the cars. The main line represents the roads (see Figure 7.33). 'Here's where you can park your lorries for two hours while you sit on the beach. You have to pay money, it's six pounds,' said Isaac. He began tearing off raffle tickets. 'I have to write how much they paid on the back; the people get through the "no entry". It's about sixty pence they have to pay.' 'We're going to have signs to say what the speed limit is,' said Daniel. Daniel and Isaac created imaginary accidents and problems with the vehicles such as crashes. Daniel's helicopter rescued the people, while Isaac stood all the people (little wooden play people) up in a crowd 'to watch'. They collected the resources they wanted to use and combined materials to make their playscape, inventing their own symbols, such as a square to stand for the gate being shut. Their persistence and excitement continued through the week and again Isaac, interested in signs, thought of his own 'opening' sign.

Daniel and Isaac were involved in high-level play that they initiated and they wove in imaginary stories and challenging problems that they solved. They were

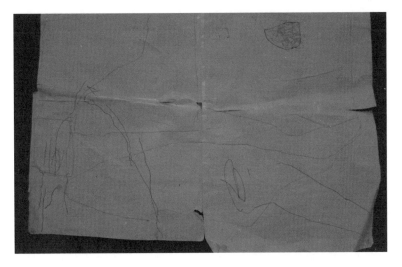

Figure 7.33 Map of the car park at the beach and road

totally engrossed as they organized the plan carefully, using any materials they needed to further develop their play and narratives. The culture of the setting encouraged them to return to their theme over several weeks. They were bringing everything they knew about beaches, car parks and helicopters to their play and were completely free to invent the symbols they needed.

Conclusion

In these episodes children are in an environment where they can play freely and choose materials. Importantly they have time to play and to return to their play many times if they need to. Having a rich, child-initiated play environment encourages children to engage in both short, sustained and extended periods of play.

In Chapter 8 the case studies show the value of collaborative play as children explore their *mathematical graphics* in group contexts.

Reflections

- When you observe children's role play, note the influence of new media, new technologies and popular culture on their play. How do these interests show themselves in the children's words and actions and in the artefacts they use or make? Discuss the various influences on the children's pretend play with you colleagues.
- Collect several examples of children's mathematical graphics and with your colleagues discuss how the children used their graphics to communicate their mathematical meanings.

8

Children's mathematical graphics in small group contexts

'We should of made a trap door' Joseph says. 'Let's just pretend *there's a trap door'*
I suggest. 'No. It has o be a real one. I'll make you one.' He takes a large piece of easel
paper and begins to draw a maze of lines and arrows. 'Put the lock part over here,'
Alex directs.'

(Paley 1990: 46–47)

What is this chapter about?

- **Small group contexts:** playing, representing and communicating mathematical ideas.
- **Sustained enquiry.**

The case studies in this chapter concern children who have chosen to play together with their peers, or whose teachers have *planned* a small group to provide opportunities for children to represent their mathematical thinking, if they choose to do so.

Small group contexts: playing, representing and communicating mathematical ideas

CASE STUDY

Measuring tools

Children

Oliver and Peter (3 years, 8 months) and Isabella (4 years, 4 months)

Teacher

Philippa

Context

Free play. Philippa had noticed the children's interest in drawing lines and set out a large sheet of paper, pens, rulers and set-squares

The mathematics

Measurement, shape and space

Dimension of the taxonomy

Written number and quantities: *early explorations with marks – attaching mathematical meanings; representing quantities that are not counted*

Isabella started drawing a train, then, discovering she could lift the flap of paper on which she had drawn, decided to draw 'another one the same under the flap'. She concentrated as she drew, showing an eye for arrangements and shapes. Peter used a ruler to make a triangle and then used the set-square as a template, pivoting it to get the shape he wanted. Oliver used the rulers to measure the length of the table, lining them up carefully in a straight line and then used them to draw across the table. Other children also added to the graphics, one drawing around her hand and another making a range of marks including straight lines (see Figures 8.1–8.6).

Figure 8.1 Exploring measuring tools (i)

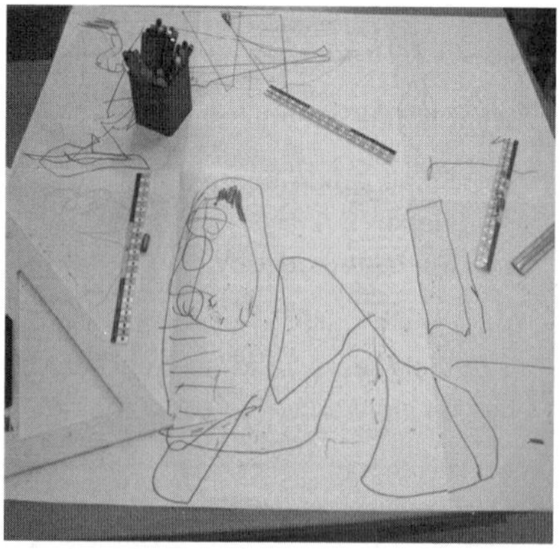

Figure 8.2 Exploring measuring tools (ii)

Figure 8.3 Exploring measuring tools (iii)

Figure 8.4 Exploring measuring tools (iv)

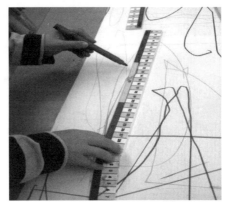

Figure 8.5 Exploring measuring tools (v)   **Figure 8.6** Exploring measuring tools (vi)

The children used a great deal of mathematical language as they chatted about their lines, shapes and drawings, developing their understanding of measurement and shape. The free play context, combined with the real mathematical resources, encouraged the children to explore aspects of measurement in open-ended ways.

CASE STUDY

Inventing shapes

Children

Nicholas (5 years, 5 months) and a group of children (5 years of age)

Teacher

Emma H.

Context

One of the children had remarked that 'Shapes are made with one line' and Emma had asked if they could show (prove) that this was so

The mathematics

Shape and space

Dimension of the taxonomy

Written number and quantities: *representing quantities that are counted*

Nicholas got a whiteboard and started drawing, then decided he wanted 'spiky bits' on his shape, commenting, 'I know it's got more than one side, but I like it

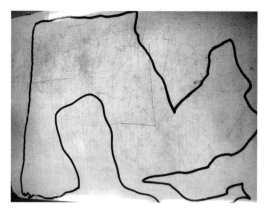

Figure 8.7 Nicholas' 'shape with one line'

Figure 8.8 Sharing understandings about 2D shapes

better now – it looks kind of scary' (see Figure 8.7). Emma asked him to explain and, laughing, said, 'It's like squares and those long squares gone wrong.' This led to a discussion about whether a 2D shape *could* only have one side and resulted in agreement that they were referring to shapes drawn with one line but which had to have '*more than* one side' (i.e. enclosed spaces). The children represented their own invented shapes (see Figure 8.8), discussing, 'What kind of lines does it have?' 'How many sides has it got?' Toby and Arushi came to the conclusion that a shape can be as big or as small as you want but has to have 'Curvy, wibbly wobbly or round bits.' Jonathan went a step further and said, 'If it has points or spikes then it has more sides than one.' This led to many shapes being drawn in the sand and on the patio with the chalks and rich discussions about 'sides', 'lines' and 'corners'. Several months later they revisited this interest, deciding to draw a many-sided shape which ended up with 40 sides – they thought this was really amazing! This collaborative dialogue with their teacher enabled the children to co-construct and negotiate understandings about shapes.

Open-ended approaches to teaching shape, space and measurement can be of enormous value to young children, allowing them to explore and make meanings about the properties of shapes and the use of measuring tools. As these two examples show, they can also provide meaningful contexts for children to use mathematical vocabulary.

Interestingly, we have observed that when younger children of 3 and 4 invent their own shapes they sometimes name them, using descriptive language that they relate to objects they are familiar with, such as 'pointy sun' (for a circle with a triangle attached), 'wiggly rocket' (for a curved shape with a pointed section at the top) or 'long eye' (for an ovoid with a circle inside it). In contrast we've found that children of 5 and above often incorporate specific mathematical vocabulary that they know into the 'names' they give their invented shapes, such as 'parallel roads' for two narrow horizontal rectangles, 'feet with right-angles' for 'L' shapes or 'pyramid hats' for equilateral triangles on an oblong base.

CASE STUDY

Plane tickets

Children

A group of 3- and 4-year-olds in a nursery

Teacher

Julie

Context

The children were keen on planes and often lined up chairs to make several rows of seats inside the plane

The mathematics

Counting, writing numbers

Dimension of the taxonomy

Written number and quantities: *early explorations with marks – attaching mathematical meanings; exploring symbols; early written numerals; numerals as labels; representing quantities that are not counted; representing quantities that are counted*

The 'driver' always sat in the front and the children would get on and off the plane as if they were travelling on a bus. Julie decided to make some 'tickets' and the children used them to write their own seat numbers for the plane (see Figures 8.9, 8.10).

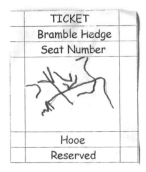 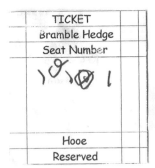 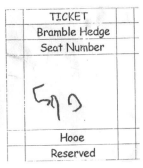

Figure 8.9 Connor: 'number 13'; Jonathon: '3' and Sam: '8'

Abigail (Figure 8.11) used a number she knew and could write (she is 4) to signify the number she wanted ('8'). She also represented the number 5 on another ticket (see Figure 8.12).

Figure 8.10 Louis: '89'; Sarah: '7' and Joe: '7'

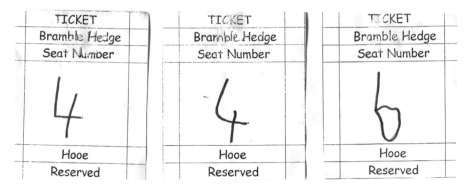

Figure 8.11 Sarah: '4'; Abigail: 'number 8' and Elliot: '6'

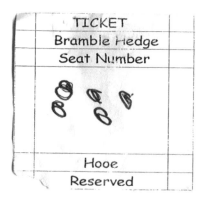

Figure 8.12 Abigail: 'number 5'

The nursery staff had reflected on the children's interest in planes, and their idea to produce 'tickets' encouraged the children to represent numerals and quantities in a variety of ways that made personal sense, capturing the children's symbolic representations in that particular context on that day. Their play led to the children choosing to use tickets in a variety of other situations.

CASE STUDY

Holes

Child

Ross (3 years, 3 months)

Teacher

Emma B.

Context

Group book time

The mathematics

Space and shape, size, matching

Dimension of the taxonomy

Written number and quantities: *quantities that are **not** counted*

After hearing the story *A New House for Mouse*, Ross drew on the whiteboard, representing one hole at a time, with marks inside each (see Figure 8.13). As he drew each circle he told his key person about his graphics, labelling them as 'A hole for a chicken to live; a monster hole; a snake hole; a baby snake hole; a dragon hole; a big dinosaur hole.' Ross focused on the relative sizes of the holes and carefully matched each to the size of the animal.

Figure 8.13 'A hole for a chicken to live, a hole for a mouse . . .'

The story had captured Ross' imagination. He used the whiteboard and marker pens available, and showed that he was aware of the size of each animal and was able to represent this. Not surprisingly, a story will often spark off a whole range of thinking explored through graphics.

CASE STUDY

Brandon's registers
Child

Brandon (4 years, 1 month)

Teacher

Emma B.

Context

At the beginning of the day

The mathematics

Data-collecting

Dimension of the taxonomy

Written number and quantities: *early explorations with marks – attaching mathematical meanings; explorations with symbols; early written numerals*

Brandon filled in a register form of his group, recognizing his friends' names and making circular and linear marks to indicate their attendance (see Figure 8.14).

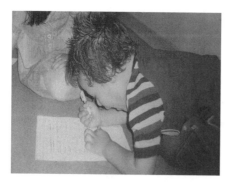

Figure 8.14 Brandon's first register

Figure 8.15 Brandon represents his friends on the whiteboard

After 15 minutes he said, 'Everyone's here.' Later he asked to use the whiteboard and represented his friends' names on it, labelling each mark and symbol (and mostly using the initial letter of their names): Brandon, Finn, Aimie and Harry (see Figure 8.15).

Brandon used a range of symbols and found a meaningful 'shorthand' method to record attendance, using two different surfaces on which to communicate his ideas. Register time is very meaningful for children and one that they constantly want to represent. Having the appropriate materials at hand gives children an opportunity to try out their marks and symbols.

CASE STUDY

The racing marbles game

Child

Finnian (4 years, 3 months)

Teacher

Emma B.

Context

Group-time marble game

The mathematics

Counting and data

Dimension of the taxonomy

Written number and quantities: *exploring symbols, representing quantities that are counted*

During group time the children had initiated a game racing marbles down a slope. Finnian watched and decided to create a scoreboard. 'I'll write the scores,' he said. He asked his key person to write at the top of the scoreboard the names of the children who were playing the game and made a vertical mark when each group scored a point. After several turns Finnian realized that Aimie's team was the clear winner with 10 marks and Harry's team had 7. At first Finnian used a line across the tallies and then abandoned that idea. He decided to draw a circle around each set of marks and linked them in the middle, using his own symbol to show that both teams were equal. 'Now I've made both teams the winner' (see Figure 8.16). Perhaps, as Harry was his friend, he did not want him to lose.

Figure 8.16 Finnian scoring points, then making everything equal

Finnian used a range of graphics to represent the scores, including vertical lines (tallies) for each person who scored, a horizontal line to link the tallies together and a circle around each score. To solve his problem of making the scores equal he tried out more than one idea, eventually drawing circles round each score. What was interesting was his emotional attachment to his friends and his sense of what is fair, rather than seeing the scores as the end result of winning or losing. Gifford (2005) also reports on children's sense of fairness in mathematics in terms of social justice.

CASE STUDY

Madison's clock

Child

Madison (4 years, 1 month)

Teacher

Stephanie

Context

At talk time with a group of children exploring a tray of clocks

The mathematics

Time, numerals as labels

Dimension of the taxonomy

Written number and quantities: *early written numerals; numerals as labels*

Madison was exploring clocks of different sizes, moving the hands of each around its face. Taking a piece of paper she wrote her name across the top, then drew a circle and added hands to her clock, explaining, 'It's seventeen o'clock' (see Figure 8.17).

Figure 8.17 Madison: 'It's seventeen o'clock'

Madison had drawn some letter- and number-like symbols and the clock's hands, explaining her thinking by telling her clock's time, as she explored her understanding of time and number.

CASE STUDY

Growing a beanstalk

Children

Mya, Danny, Oliwier, Amelia and Isaac (4–5 years)

Teacher

Janet

Context

Janet had read the story of *Jack and the Beanstalk*. On this occasion she purposefully put out a tub of soil with eight flowerpots, a fully-grown bean plant and 14 beans to encourage the children to think about how they would share the beans equally

The mathematics

Counting, quantities, estimation, division by sharing, fractions, remainders, time

Dimension of the taxonomy

Written number and quantities: *numbers as labels, representing quantities that are not counted, representing quantities that **are** counted*

Figure 8.18 Planting beans

The dialogue developed as the children thought about how long the beans would take to grow, as Janet set about sharing the 14 beans between five children (see Figure 8.18):

'Growing a beanstalk?'
'How long does it take to grow?'
'Eighty, a hundred nine sixty days.'
'Eighty-six.'
'Eighty, seventy, a hundred.'
'Sixty, eighty days.'
'Five days.'
'I'm going to hold the seeds.'
'How get soil in pots?'
'I know. Soil in pots – use hands . . . then finger in, make a hole . . . put seed in and water.'
'How?'
'How many pieces of soil?'
'Count it. I count pots. Count seeds Oliwier. Amelia count children.'
'One, two, three, four, five, six . . . fourteen seeds.'
'Five children.'
'Got too many.'

'Grab soil till right to top, pat it down and spread.'

'I need to count one, two, three, four, five, six. No actually, two left in bag . . .
   eight.'

'I got another hole – two each.'

'One for you, for you . . .'

'Has everyone got seeds?'

'Three left.'

'We could break them up.'

'Cut them up – we need to cut in half.'

Olivier cuts two seeds in half and gives some of the halves to his friends.

'We don't need to cut this one . . . I can have it.'

'Need to water them.'

'Fill it up.'

Two of the children chose to use paper to support their thinking about the problem. Isaac explained, 'That means lots of seeds cutting up' (see Figure 8.19).

Amelia (Figure 8.20) wrote '41' (for the 14 beans) and above it wrote '5' for the number of children who were planting beans. She decided to include one more bean for her friend who was ill and revised the number of children to reflect this: her calculation was now based on this new total of the five children in the group, with the addition of her friend. Next (forgetting the original number of five children) she worked out mentally that if she gave each of the six children one bean there would be eight beans remaining, and wrote '8' on the right. Later she returned to her graphics and, thinking about sharing the 14 beans equally between the five children in the group, worked out that each could have one of the eight remaining beans. Finding that this would leave just three beans to share (written next to the '8') she drew these, showing how she would divide them in half. Amelia's graphics are a good example of how children use their mathematical graphics to support their thinking as they solve problems, although in this instance Amelia's decision to change the number of children in her problem confused her a little.

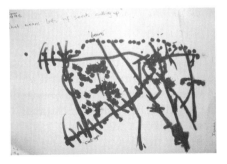

Figure 8.19 Isaac: 'Lots of seeds cutting up'

Figure 8.20 Amelia works out how she could share the beans

The children's collaborative dialogue appeared to be of particular value in supporting their thinking. This example suggests that occasionally making a transcript of a discussion can provide adults with valuable information about children's understandings.

CASE STUDY

One is a Snail, Ten is a Crab

Children

Tyrees, Jonny, Roxy, Harrison, Shimae, Jamie and Travell (4–5 years)

Teacher

Karen

Context

The delightful picture story book, *One is a Snail, Ten is a Crab* (Pulley Sayre and Sayre 2004) uses pictures of snails, crabs, dogs and spiders – each with different numbers of feet to support counting to 100. After sharing the book with the children, Karen suggested that they choose their own number and work out which combination of creatures' legs would total that number.

The mathematics

Counting, counting in multiples, counting on, counting back, estimation, repeated addition, subtraction, multiplication

Dimensions of the taxonomy

Written number and quantities: *numerals as labels, representing quantities that are counted; exploring symbols*

Calculations: *children's own written methods: counting continuously; separating sets; counting with larger quantities*

Shimae had decided to 'find out what is 100'. She wrote the number '100' and as she progressed, counted the legs she'd drawn (see Figure 8.21). After a while she said, 'This is hard now, all this counting. Can you help me?' So as Shimae drew, an adult counted with her, Shimae noting the total she counted each time. When she'd reached 100 she wrote: '6 spidr 3 insec 1 pursn 1 well (whale) 1 pig'.

Johnny (Figure 8.22) explained that he had decided to 'find out what 30 is' and wrote '03' on the top right of his paper. After a short while he paused to count the legs of the creature he had drawn: 'Ooops! That spider has got too many legs, it's meant to have eight!' Then he crossed out the extra legs he had

Figure 8.21 Shimae

Figure 8.22 Jonny

drawn. Jonny continued drawing, counting the legs each time. He said he had 22 now, needed '23, 24, 25, 26, 27, 28, 29 and 30. That's eight more – I know, easy! I can do another spider!' He then wrote: '2 sbidr, 1 crab, 4 snial'.

Harrison had chosen to work out which creatures to select so that their feet would total 11. He began by explaining, 'I don't know how to draw spiders, so I'm just going to do the legs.' Next he drew two additional legs, explaining, 'One person,' and counted them all: 'One, two, three, four . . . ten. Now I just need one more – it's a snail. There, that's eleven altogether' (see Figure 8.23).

Travell had decided to find out 'what is 40'. He drew lots of people and asked, 'Does that make 40?' The adult nearby said she was unsure and asked how he could find out. Travell responded by counting in twos: 'Two, four, six, eight, ten, twelve, fourteen, sixteen, eighteen, twenty.' He appeared to be stuck and turned to look at the '100 square', continuing counting in twos to 32, so that he had counted all the people he had drawn. Finally he drew four additional people and counted the total, writing: '20 people makes 40' (see Figure 8.24).

Figure 8.23 Harrison

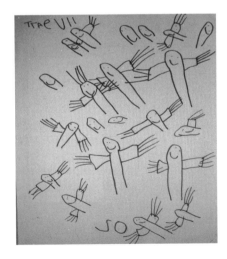

Figure 8.24 Travell

Roxy decided she was going to work out 'what is 67'. 'It's a big number so I'll do crabs because they have the most legs.' After drawing four crabs she counted, 'Ten, twenty, thirty, forty. Four crabs is forty,' and wrote this down. Drawing six more crabs she added, 'That's six crabs – ten, twenty, thirty, forty, fifty, sixty. Now I need seven more.' Roxy drew three people. 'That's two, four, six, so I need one more. That's a snail!' Finally she wrote: '6 crabs, 3 peepoll, 1 snail macs 67' at the foot of the page (see Figure 8.25).

As the teacher read the book, Jamie commented, 'I know! Nine could be an octopus and a snail.' Then he too decided to work out the combination of creatures that would total 100. 'Because that is massive!' He wrote '9 peeple', then used his fingers to count, 'Two, four, six, eight, ten, twelve, fourteen, sixteen, eighteen' and wrote: '2 spiders'. Counting on 18, he counted on eight more, then another eight. He then wrote: 'I optpus 2 spiders' and counted on eight more. 'This is getting hard now. I know! I can write it down each time and count on!' After writing what he was adding on, Jamie wrote the total. When he got near the end he added '9 spiders' then crossed it out and wrote '8', explaining, 'Nine makes too many legs – so I had to take one off to make it the right amount' (see Figure 8.26).

Figure 8.25 Roxy: 'What is sixty-seven?'

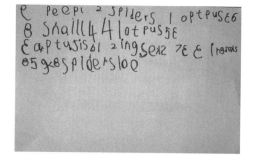

Figure 8.26 Jamie

As the book was read, Tyrees burst out, 'I know why ten is a crab, because it's got ten legs – see, one, two, three . . . ten.' Then he added, 'I know, nine could be an octopus and a snail.' He explained he was going to work out 800 and reaching for some paper wrote '800', after a while explaining that he had four crabs and four snails: 'That's ten, twenty, thirty, forty and four more – forty-four! That's not the eight hundred – I need loads more so I think I'll do more crabs 'cos they've got most legs.' He continued to draw (see Figure 8.27). 'There, I done six more crabs. That's ten, twenty, thirty, forty, fifty, sixty and forty – there's a hundred and four more snails – that's a hundred and four.' Pausing, Tyree decided, 'I don't think I'm going to do any more because it's too big. I need six more so I can do six snails, or a dog and two snails, or three people – but I'm just going to do a fly because that has six legs.'

Figure 8.27 Tyrees

Karen described this as a real 'eye opener'. It was the first time she had tried to support children's use of their own graphics to assist their mathematical thinking. She said that the children's self-challenges and their individual problem-solving were remarkable. These were able children in nursery and Reception and this session revealed that ability. Karen had chosen a stimulating book that really intrigued the children. Talking to other teachers she emphasized, 'For children to use their own mathematical graphics, give them a real challenge.'

Sustained enquiry

This section focuses on episodes of sustained enquiry where children explored and developed their interests and mathematical thinking in depth within a group context, over time. Again the case studies show the collaborative, social nature of play that allows children to co-construct and develop deep levels of understanding as they draw on their home knowledge, and highlight the adults' role in supporting their thinking.

CASE STUDY

Caravans at snack time

Children

Harry (4 years, 5 months), Finnian (4 years, 1 month) and Brandon (4 years, 5 months)

Teacher

Emma B.

Context

Conversation during snack time

The mathematics

Numerals as labels

Dimension of the taxonomy

Written number and quantities: *early written numerals; representing quantities that are counted; numerals as labels*

During snack time Harry initiated a conversation about caravans. He described his experience of staying in one, telling the children that it had wheels, a kitchen and a 'long number' on it. This stimulated Finnian to tell his friends about his experience of a caravan: his had a kitchen and sofa. Harry explained that his grandma was staying in another caravan a bit further away and that her caravan had a kitchen and a sofa. Some children had not stayed in a caravan and were unsure about what it meant. Harry decided to represent his caravan on the white-board to help these children understand: he showed the 'two wheels' and wrote the long number '12148' (see Figure 8.28).

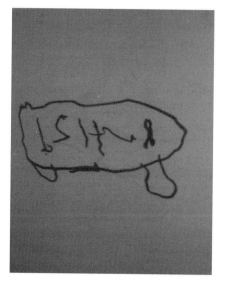

Figure 8.28 Harry's registration number on his caravan, '12148'

Finnian wanted to represent his caravan too, and he showed that there were many caravans as he drew them on the whiteboard (Figure 8.29). He explained that they were all very close together. The top caravan shows him standing on the sofa of the caravan and looking out of the window. Hs caravan also had two

Figure 8.29 Finnian's representations of his caravan and the other caravans on the caravan site

wheels. It was somewhat unclear whether these were static or towed caravans, and Emma felt that further discussion might elicit this, although the wheels suggest they could be towed.

Brandon had stayed in a caravan but told his friends that he really enjoyed travelling in his dad's car (Brandon may have classified this through his new awareness of caravans with similar properties to the car). Although his car did not have a kitchen, Brandon said that it did have seats and he liked to sit in the front. Brandon was keen to represent his car as well.

Harry and Finnian were very keen to share their knowledge of caravans through their graphics, which they carefully explained. Harry helped those children who had not seen a caravan enter the conversation by illustrating a caravan from his perspective. Numbers are significant to Harry and he recalled some of the numbers on the caravan's number plate. Finnian drew the surrounding camp-site, explaining that there were many caravans. He also included details such as the sofa that seemed to be important to him, since he represented himself standing on it, looking out of the window.

The children continued to enquire about caravans over the next two weeks and Emma brought in different photographs of caravans to show the range and variety of these vehicles and their appearance. The children's use of graphics was significant in this discussion. They demonstrated their skill in using graphics to assist their thinking and as a tool for communication. Their dialogue enabled them to explain additional details. Tizard and Hughes (1984) termed these persistent enquiries that children make, 'passages of intellectual search'. Often children do this with an adult in a one-to-one interaction, but here they collaborated to support each other's thinking.

Conclusion

Playing and exploring ideas with others provide rich social contexts for learning, enabling thinking to be shared, negotiated and co-constructed, and for children's own graphics to become peer models for others. As many of the examples in this chapter have shown, playing with others does not always need to be within an adult-planned group. Whilst we have grouped the case studies in this chapter under the heading of 'small group contexts', 'Sophie's picnic' (in Chapter 2), the 'scoring goals' case study in Chapter 6 and the case studies of sustained periods of play in Chapter 7 are all examples of rich social and collaborative (or group) play, highlighting their effectiveness.

In the next chapter we focus particularly on children using their own graphics and methods to solve mathematical problems, including the case studies of calculations and the rich collaborative dialogue of 'Is zero a number?' At the same time it is true to say that whenever children use their *mathematical graphics* they are in a sense always solving problems (e.g. *'what would best help me think about this?' 'How shall I represent the data I'm collecting, so I can effectively analyse it later?' 'How might I show the operation in this calculation?' 'What's the most effective may I can communicate my ideas to someone else?'*). Of course young children will not word questions in such explicit terms, but these questions are implicit in their thinking, and as we have argued earlier thinking and the processes of mathematics are at the heart of *children's mathematical graphics*.

In the next chapter we focus on children using their own graphics and methods to solve mathematical problems.

Reflections

- Record (transcribe) children's talk within their collaborative play. What does this tell you about their thinking and understandings?
- Looking at some of your written observations (or learning diaries) of the children's play and reflect on what they tell you about the children's own interests. How do you and your colleagues support these?
- Looking at some of your examples of children's mathematical graphics and discuss how they reflect the children's interests.
- You may like to take one of the case study examples in this chapter and discuss it in detail with your colleagues. What does the case study tell you about the children's understanding of calculations? How might you develop calculations in your class?

9

Solving problems: children's mathematical graphics

Play itself is the practicing of problems, a fact demonstrated by even the most casual attention to the passing dialogue. 'The monster is coming! He's almost here!' 'Get the magic belt! When he puts it on he gets froze!' New conflicts arise everywhere; it is our business to find the magic belt if we can.'

(Paley 1990: 80)

What is this chapter about?

- **Solving problems through graphicacy.**
- **Sustained periods of play.**

In all the case studies in this chapter we see that children are constantly solving their own problems, since this is the nature of play and children's mathematical graphics. They are more likely to engage in problems they set themselves or those they find personally challenging, whereas arbitrary problems set by a teacher and divorced from any real or classroom experiences will not have the same meaning for them. It is vital that teachers help children make connections with the complexities of the problem and the larger concepts involved (Askew *et al.* 1997). Problem-solving needs to be seen within the wider spectrum of the class environment within which teachers build a culture of mathematical enquiry.

Solving problems through graphicacy

Solving problems is at the heart of babies' and infants' lives. It is a cognitive skill that has evolved throughout the course of human development and it develops throughout the span of our lives. Problem-solving is closely related to curiosity, risk-taking, creativity and inventiveness, and rather than being a lone activity, its success is often realized though joint social activity, dialogue, collaboration and co-construction of meanings. Pound (2006: 9) emphasizes that solving problems or 'problem *finding*' is something humans seem to need to do: 'Solving problems lies at the heart of mathematics . . . children need to solve problems to become problem solvers'. Problem-solving is at the *very centre* of the graphics children choose to use to support their mathematical thinking. Gifford (2005: 150) argues that 'Challenge and choice of method are therefore key characteristics of problems. If children already know or are told the method to use, then they are not problem solving'. In England, as the curriculum area of '*problem solving, reasoning and numeracy*' suggests, solving problems is also central to mathematics (DfES 2007a).

Through several case studies, this chapter focuses on children's individual mathematical problems and the strategies they used to help them reach a solution. It considers the role of problem-solving in the EYFS, focusing on the guidance for mathematics and showing that the different interpretation of 'problem-solving' as a largely adult-set mathematics problem must not result in 'top-down' pressure and mathematical 'problems' that only belong to the adult. We also explore calculations as children investigate personal strategies through their own *mathematical graphics* to help them solve problems in personally meaningful ways.

In England, the Williams Maths Review argued that: 'Issues [difficulties] relating to '*using developing mathematical ideas* and methods to solve practical problems' relate in part to practitioners not recognizing this or not providing an environment where this can take place, rather than reflecting on most children's inherent capabilities' (DCSF 2008a: 36).

CASE STUDY

Lily and the whiteboards

Child

Lily (5 years, 0 months)

Teacher

Jan

Context

Jan was busy working with a group when the school secretary delivered a parcel of 20 small whiteboards

The mathematics

Division – sharing a quantity equally between three (and showing a remainder)

Dimension of the taxonomy

Calculations – children's own written methods: *counting continuously; separating sets; exploring symbols; calculating with larger numbers*

Lily offered to sort out the whiteboards, to share them equally between the three Reception classes so that she would know how many to give to each class. She began dividing the whiteboards into three piles (adding one board to each pile in

Figure 9.1 Lily and the whiteboards

turn). As she did so, she wrote a cross for each on a scrap of paper, then, drawing the three teachers on a larger piece of paper, she wrote their names, and recounted the boards, writing each numeral in turn as she did so (see Figure 9.1). Lily checked her progress, beginning with the six boards for one of the teachers (on the right) by ticking each and writing 'Done'. At one point she noted (on the left) 'Need to do' for another teacher and then added a cross as she checked each board in another pile. She then wrote 'Need now' as she continued. Lily twice wrote 'No' with a cross, as she identified the two boards that remained after she had equally shared all 20 boards. Finally Lily put one pile of six whiteboards in the graphics area of her own class and delivered the other two piles of six to each of the other teachers. On her return she gave the two surplus boards to her teacher. Expecting Lily to just give her a piece of paper with a number, Jan was surprised when Lily gave her a paper with an illustration, showing the method she had used to solve the problem.

In this example Lily's drawings of the three teachers were not vital to the mathematics, although they may have initially helped her focus on the problem. It is important for teachers to use their knowledge of individual children to help scaffold the children's leaning through collaborative discussions: these will not only help children think about the meanings of their symbols and representations, but will also help them move to increasingly efficient methods over time. Jan said that Lily often instigated her own projects and problems and that she encouraged her to do this. Small groups of children can also benefit from solving problems such as this, since they allow children to share and discuss their various strategies and negotiate meanings.

CASE STUDY

Maisie's measure

Child

Maisie (4 years, 7 months)

Teacher

Julie

Context

Late in the autumn term the children in Maisie's class had been making paper-chains to decorate their classroom

The mathematics

Measuring length; comparison

Dimensions of the taxonomy

Written number and quantities: *representing quantities that are counted; numerals as labels*

Calculations – *children's own written methods: counting continuously*

Maisie decided to see how long her paperchain was. Although there were several 30cm rulers on her table she decided to make her own 'ruler', and writing as many numbers as she could fit on a strip of paper she stuck on another strip when she reached its end. She continued writing numbers on the second strip from where she had left off on the first, and as she added to the length of her paper ruler she checked it was sufficiently long and then measured her paperchain, finding it was '27 long' (see Figure 9.2).

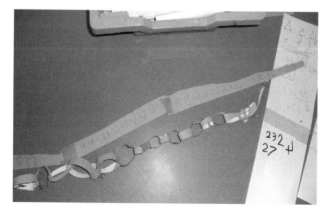

Figure 9.2 Maisie's measure

Maisie's teacher, Julie, encourages open thinking and ways of working and, since this was a problem that Maisie posed it was real and meaningful to her, supporting her growing understanding of linear measurement and the function of rulers.

CASE STUDY

Stanley's shoes

Children

Hannah (5 years, 5 months) and Stanley (5 years, 6 months)

Teacher

Kylie

Context

There were nine children in the home corner who had taken off their shoes during their play and at tidy-up time Kylie asked, 'How many shoes will we need to find?', suggesting they might find a way of working this out

The mathematics

Writing numbers, counting

Dimension of the taxonomy

Written number and quantities: *numerals as labels, representing quantities that **are** counted*

Figure 9.3 Hannah counts all the legs

Figure 9.4 Stanley: counting the pairs of shoes

Hannah decided to fetch some paper and drew nine people, then counted all the legs she had drawn and wrote '18' at the foot of the paper (see Figure 9.3). Stanley drew a horseshoe shape to represent each pair of legs and feet. He then counted in twos, writing each number down beneath a horseshoe (Figure 9.4). He had missed counting the 'feet' of one horseshoe and at first counted 16 (the children had been learning and counting things in twos the week before). When they looked together at their different methods, Stanley quickly realized what he had done and counted again, this time counting them all to get the total.

CASE STUDY

Counting tentacles

Child

Lucas (4 years, 6 months)

Teacher

Philippa

Context

Lucas had brought a toy with numerous pink tentacles into school and this sparked his question: 'I wonder if there are more tentacles than my mum's age?'

The mathematics

Counting; estimating; grouping in sets of 10; counting in 10s; calculating a total

Dimensions of the taxonomy

Written number and quantities: *representing quantities that **are** counted*

Calculations – children's own methods: *counting continuously; separating sets; calculating with larger numbers*

At first Lucas tried counting each tentacle but soon realized that it was too difficult to keep track of those he had already counted. He asked a friend to help hold some of the tentacles and finding this difficult, he decided to look for a physical tool and used some small paper clips he found. Philippa noticed that the paper clips were small and offered some larger 'bulldog' clips that were easier to use, asking him how many tentacles he was putting in each. This question encouraged Lucas to reconsider his previous random groupings and he decided to group them in 10s, a number that could be satisfactorily held in each bulldog clip (see Figure 9.5). As he made groups of 10 Lucas took a piece of paper that was nearby, noting how many sets of 10 he had (Figure 9.6), though abandoning this when beyond 80. When all the tentacles were held in bulldog clips Lucas counted how many he had altogether: he had a total of 12 groups of 10 – certainly a lot more than his mum's age!

Figure 9.5 Lucas counting tentacles

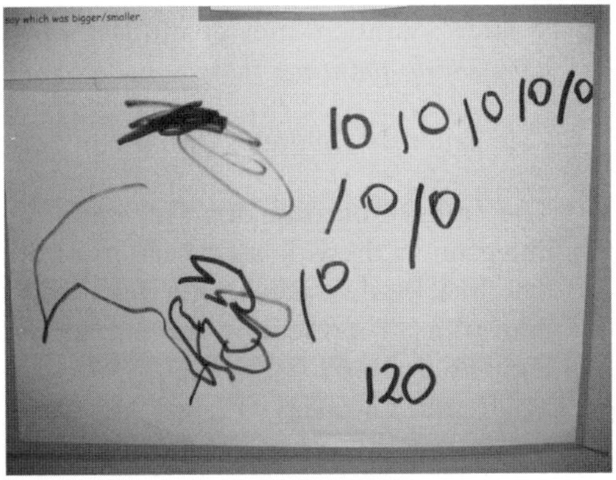

Figure 9.6 Lucas keeping track of the sets of tentacles

This example originated through this child's curiosity about his toy (and his mum's age) and began with a problem he had posed. Philippa's encouragement for Lucas to solve his problem enabled him to explore various strategies and their joint discussion about the quantity of tentacles held in each bulldog clip scaffolded his thinking further, enabling him to reach a final answer.

CASE STUDY

How many drinks?

Children

Harvey, Remee, Macaley and Leo (4–5 years)

Teacher

Becky

Context

For their trip to the woods, the children knew that they would need to take two drinks each. Before they left their teacher provided cartons of drinks and asked if they could help her work out how many cartons they would need to take altogether, for the group of 10 children

The mathematics

Counting

Dimensions of the taxonomy

Written number and quantities: *early written numerals; numerals as labels; representing quantities that are not counted; representing quantities that **are** counted*

Calculations – children's own methods: *continuous counting; separating sets*

The range of thinking is important. Harvey represented people and interestingly drew dots for cartons of drink (see Figure 9.7). He obviously knew that this would take a long time and that it would be more economical to draw dots, representing *quantities that are not counted* (see the taxonomy, Figure 6.8, p. 76).

Remee's strategy (Figure 9.8) was to draw 10 children and then draw two cartons for each child. It appears she squeezed the last three children into her drawings, and that it was difficult to draw the cartons beside them. This meant that Remee was two cartons short, but overall she had an effective and very workable strategy.

Leo used the same strategy as Remee and counted the cartons, putting 20 on the right-hand corner of the paper, then wrote the number 10 for the 10 people he represented (Figure 9.9).

Macaley used numerals (Figure 9.10) and while we do not know what else he was thinking, it is clear that the '2' was significant to him.

Figure 9.7 Harvey drew 10 children – he explained that the dots represented 'lots of drinks'

Figure 9.8 Remee also drew 10 children and 18 drinks

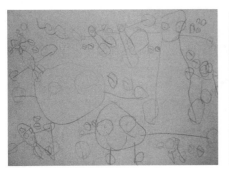

Figure 9.9 Leo shared out the drinks, giving two to each child

Figure 9.10 Macaley wrote lots of '2s' to help him think about the many drinks they would need

CASE STUDY

Biscuits for bears

Children

Callum, Luke and Zainab (all 4–5)

Teacher

Sara

Context

The children were having a 'teddy bears' picnic' and there were some surplus biscuits. Sara asked the children how they might share four biscuits equally between the six children

The mathematics

Counting, division by sharing, fractions

Dimensions of the taxonomy

Written number and quantities: *representing quantities that are counted; numerals as labels*

Calculations – children's own methods: *counting continuously; separating sets*

Zainab (Figure 9.11) had decided to share six biscuits between four children. She wrote numbers 1 to 6 and then drew four circle shapes to represent biscuits. She drew a line from each number to a biscuit and had two left over. Zainab did not know what to do with the remaining two so she divided them into quarters,

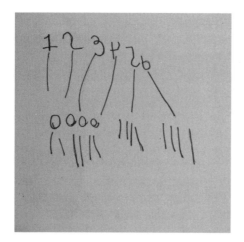

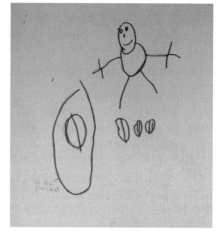

Figure 9.11 Zainab   **Figure 9.12** Luke

drawing four lines to represent the four quarters. Zainab's graphics are clear and organized and, although she did not completely solve her problem, she was thinking about fractions as she split the remainder.

Luke (Figure 9.12) drew one person to represent the six children and then drew four biscuits. He decided he would cut the biscuits in half, sharing out six halves one to each of the six people. However, there was still one remaining biscuit that he said he would put in his back pocket! He put a circle around the remaining biscuit. Like Zainab, Luke did not entirely solve his problem but organized his thinking very clearly and highlighted the one left over.

Zainab reframed the question: she may have thought that sharing four biscuits between six people was too difficult to work out and that adjusting the question would offer an easier problem. Luke found the problem difficult but proposed a sensible solution, as children often do. Both Zainab and Luke offered part solutions.

It is really important to look at the processes involved in children's thinking and not get too focused on the correct answer. By valuing the children's own methods and really listening, teachers can uncover what they know (rather than what they do not know). The mathematical graphics also help the teacher to 'see' their thinking.

The problem in this case study was a challenge since although the numbers involved were small, they could not be easily divided. It provided a natural opportunity for discussion about remainders and fractions. The children had some understanding of fractions, revealing that the next steps for the teacher could be to model different ways to tackle fractions, without the children copying the teacher's model and allowing them to think about it themselves (see Chapter 10).

Sustained periods of play

Voldemort and the spiders' legs

Children

India, Finn, Arthur, Jo and Catherine (4–5 years)

Teachers

Donna, Jo and Keeley

Context

With the release of the latest *Harry Potter* film, the children had been talking excitedly about various aspects of the story's plot and characters. One of the children mentioned spiders having their legs pulled off and the teacher used this as an opportunity for subtraction

The mathematics

Subtraction as 'taking away'

Dimension of the taxonomy

Calculations – children's own written methods:

India: *separating sets; exploring symbols and standard symbolic operations with small numbers*
Finn: *counting continuously; separating sets; exploring symbols*
Arthur: *counting continuously; separating sets and standard symbolic operations*
Jo: *counting continuously; separating sets; exploring symbols and standard symbolic operations with small numbers*
Catherine: *counting continuously; exploring symbols*

In Figure 9.13 India used both horizontal calculations and drawings of some of the characters including Voldemort stealing spiders' legs. She used both standard operands '–' and '=', crossing out (at the foot of the page), and a box to signify subtraction. In Figure 9.14 Finn subtracted the spiders' legs by scribbling over each leg of four of the spiders he had drawn, explaining that they were now 'Naked spiders who've had their legs taken away.'

Arthur began by drawing two spiders but then became engrossed in seeing how far he could count and how many of those numbers he could write (Figure 9.15). The children's excitement about their mathematics led other children in the coming weeks to explore subtraction in different contexts. For example, Jo (Figure 9.16) and Catherine (Figure 9.17) combined earlier

Figure 9.13 India

Figure 9.14 Finn

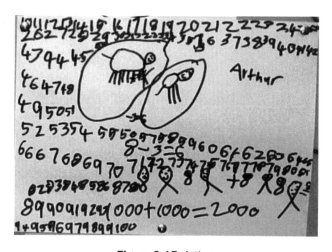

Figure 9.15 Arthur

Figure 9.16 Jo 'stealing fruit'

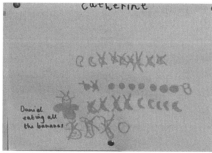

Figure 9.17 Catherine: 'Daniel eating all the bananas'

strategies they'd used when subtracting spiders' legs with a discussion about fruit at snack time.

As Figures 9.16 and 9.17 show, other children drew on the personal strategies their peers had used earlier when subtracting spiders' legs, demonstrating the significance of peer modelling in children's mathematical graphics. The children's interest led to a display of the different strategies they had used to work out their calculations (see Figure 9.18) and showing how teachers can sometimes tune into children's interests (such as the *Harry Potter* stories) and apply them to calculations.

Figure 9.18 Display of subtraction, including 'Voldemort and the spiders' legs'

CASE STUDY

'Is zero a number?'

Children

A group of 3- and 4-year-olds in a nursery at a children's centre

Teacher

Carole

Context:

A 'quiet time' where children could engage with each other and their key person. A range of resources was available, including pens and paper, books, games, magnetic numbers and a whiteboard. This was the period after lunch when the children could rest and relax and choose their own resources

The mathematics

Discussion about digits and numbers, place of digits within a number, place value

Dimension of the taxonomy

Written number and quantities: *using numerals as labels*

The children were engaging in finding out about numbers through placing magnetic numerals on a whiteboard, which could support their understanding of written numerals. Max explored the basket of magnetic numerals and placed a line of number 9s onto the whiteboard (see Figure 9.19). Daisy, noticing what Max had done, came to join him and gave him other number 9s that she had found around the room and in the basket. She then placed a 0 onto the board. Felicity, pausing while drawing at a nearby table, looked over and commented: 'My mummy does not think zero is a number.' Alfie explained, 'Zero *is* in number ten.' The nursery teacher picked out a 1 and a 0 from the basket of magnetic numerals and Alfie continued: 'Put another zero and that is one hundred.' Jack had been listening and said, 'My mummy says one, two, three, four, five, six, seven, eight, nine, ten' (he counted on his fingers as he did this). Jack then picked up the magnetic 1 and 0 and stuck them on to the board, saying, 'Look Carole, see what I've done.' Felicity, still watching, said, 'Five, six and one makes . . .' Carole wrote the numbers on the board and repeated them back to her, then added them up, putting the total, which she then read out to Felicity. Mason came over and said, 'My favourite number is number five. When I get bigger . . .' Daisy continued, 'My favourite number is one hundred.'

Alfie, obviously thinking about how you make these numbers, said, 'If you put a five and a zero that's fifteen.' Carole said, 'That's really good thinking, fifteen looks like this,' and she placed a magnetic 1 and 5 on the board. Alfie thought about this and then said, 'I mean fifty', self-correcting. 'And if you put a six in it, it will be sixty.'

At this point Daisy was becoming increasingly interested in what number could be made if you put certain numerals together (see Figure 9.20). She tested

Figure 9.19 Max placing magnetic numerals on the whiteboard

Figure 9.20 Daisy asking what numbers the combined digits represented

this out by sticking a group of numerals on the board. She asked, 'What number is this?' Carole answered, 'Sixty-three thousand and one.' Daisy then proceeded to move the 0s from the 63,000 and 1 over to the other side of the board, asking each time, 'What number does this make?' Carole told her what the number was each time she moved the numerals: 631, then 63, then 3 and then 0.

This was a very rich episode of a group of very young children collaboratively thinking about numerals and asking questions. It was a spontaneous example of mathematical thinking and one that Carole capitalized on and facilitated. She explained how she knew that this was 'a nugget, a golden opportunity to support and nurture the children's thinking as far as they would go'. The children listened very carefully to each other and picked up threads of conversation to take whichever way they wanted. When children do more talking than the teacher we can be sure that they are actively thinking and following their own line of enquiry. This is a 'master teacher' at work who is on the inside of children's learning (see Chapter 10).

Taking this episode forward, over the next few weeks Carole displayed larger numbers on the number line and put three-digit numbers on the bikes – for example, 236, 105 and 342. This extended the children's talk as they asked 'What number is my bike?'

Conclusion

In many respects all of *children's mathematical graphics* involve children in solving problems: they need to decide how to represent their thinking and how to communicate that thinking to others. When calculating they need to consider which symbols and layout will help them calculate, and which methods to choose that are relevant to the problem's context and the operation.

In the next chapter we reflect on issues of agency and identity and how they relate to young children's graphicacy.

Reflections

- Gather examples of children's mathematical graphics that they have used to help solve problems, both from their self-initiated play and from adult-led small-group contexts. Discuss the various strategies the children used to help them solve each problem.
- Using the taxonomy, compare and evaluate (from a positive perspective) the different strategies that a group of several children used to solve the same problem: what do these examples tell you about the different children's mathematical understandings?

SECTION 2

Pedagogy and practice

10

Children's voices

Jason has his own design for learning ... He presents a vivid image, but how do I mark his growth? ... As I learn to listen to what he tells us about his helicopter fantasy, I begin to see in new ways that only by reaching into the endemic imagery of each child can we proceed together in any mutual enterprise. All else is superficial; we will not have touched one another.

(Paley 1991: 12)

What is this chapter about?

- **Agency and identity**.
- **Democratic learning cultures**.
- **Listening to children**.

The right for children's voices to be heard and to have their opinions taken into account was first highlighted in the United Nations' *Convention on the Rights of the Child*: 'The child has the right to the freedom of expression. This right shall include freedom to seek, receive, and impart ideas of all kinds, regardless of frontiers, either orally or in writing or in print, or in the form of art, or through any other media of the child's choice' (UNICEF 2009, Article 13).

In this chapter we raise questions about the ways in which children's drawings, writing and their *mathematical graphics* may be perceived differently in the contrasting culture of nursery and home and argue that by listening and hearing children's 'voices' we can best support their developing sense of personal agency and identity. This is linked to the need to understand and assess children's play and their mathematical graphics from a positive perspective.

In England the document *Every Child Matters* (DfES 2004) underpins the EYFS (DfES 2007a) and emphasizes the importance of 'listening to the child's voice'. A more recent publication (DCSF 2010) emphasizes the significance of 'building on children's interests' – though this is important for all children and not only those who are viewed as 'gifted and talented'. Our argument is that observing, recognizing and acknowledging children's individual interests, contributes to rich (and often sustained) episodes of play.

Listening to children's voices means listening not just to their opinions and views: children 'speak' via a range of personal expressions including drawing, imaginative play, models, dance and narrative (e.g. Lancaster and Broadbent 2003). Our interpretations of these voices relate to children as central players (in their learning and play) and support the concept of children's 'voices' as they explore their mathematical ideas through their marks and representations.

Children's drawings, writing and mathematical graphics combine with culture to inform different symbol systems and shape their narratives, emphasizing an alternative construct of all children as capable, intelligent and powerful agents of their own learning. This perspective challenges established perceptions of imagination, play and graphicacy in early childhood and can inform and enhance pedagogies of play and learning. It rests on a poststructural view of the child as a 'co-constructor of knowledge, culture and identity', assuming 'that children are knowers of their worlds and that, therefore, their perspectives and understandings can provide valuable insights' (Janzen 2008: 291–2). It provides an image of the child that changes in 'a pedagogy of listening' (Dahlberg et al. 2007: 102). This perspective enables children to develop what Bourdieu (1992) termed 'cultural capital' and has implications for early childhood pedagogy. Moss (2007: 7) proposes that 'democracy creates the possibility for diversity to flourish. By so doing, it offers the best environment for the production of new thinking and new practice'.

We recognize that there can often be a gap between traditional perceptions of drawing, writing and 'written' mathematics in school and children's own graphics. The implication of this gap is that it limits children's feelings of agency and identity as mathematicians and the way in which they view themselves in this respect.

Agency and identity

A poststructural perspective of children's learning supports their self-identity and independence. Inden emphasizes that agency:

> is the power of people to act purposefully and reflectively, in more or less complex relationships with one another, to reiterate and remake the world in which they live in circumstances where they may consider different courses of action possible and desirable, though not necessarily from the same point of view.
>
> (quoted in Holland et al. 1998: 42)

Edmiston (2008: 129) proposes that people 'have agency when they intend their actions, reflectively make sense of events, and are aware that choices make a difference to themselves and others'.

In education, 'agency' suggests the capacity for children to make personal choices, to have some control and to be able to shape their learning. In the context of graphicacy the implication is that children should have opportunitties to make and explore personal meanings through 'selectively using the cultural resources available to them to address whatever concerns or problems they face ... their cultural resources include *all the narratives they have explored and made their own*' (Edmiston 2008: 21, emphasis added). We can see how significant this is for all aspects of imaginative play and graphicacy, and partcularly so in respect of *children's mathematical graphics*. Rather than copying or completing adults' versions of written mathematics, or drawing and writing what they are told, children make conscious decisions about their own mathematical ideas and ways of communicating them. Van Oers (2005: 7–8) proposes that imagination 'is a basic element for individual empowerment within play and for the development of critical cultural agency' that promotes development. In turn, children's personal agency supports their feelings of self-worth and contributes to their beliefs and self-identity, enabling them to see themselves as capable and confident – particularly significant in the realm of learning mathematics.

Identity is seen as a 'complex and changeable construct, influenced by socio-cultural factors' (Hall 2010: 96). Drawing on work by Coté and Levine (2002), Hall (2010) acknowledges that personal identity is either 'affirmed or discredited by the individual as well as being either validated or challenged by others', adding that (according to de Ruyter and Conroy 2002) adults play a significant part. Drawing, 'when it combines everyday experiences with imagination – also offers an authoring space for self' (Hall 2010: 97). Regarded by Moyles (1989: 12) as 'a type of intellec-tual play', drawing allows children to create their own worlds and cultures (Thompson 1999). Indeed, when children explore their ideas though imagination and play, their meanings will contain traces of their identities, and are infused with the children's personal identities, linking personal, home, community and global cultures with contemporary popular cultures, media and technologies. It is the child's ownership of their graphics, the child's personal 'voice' that is directly linked to their personal iden-tity and, as Pahl and Rowsell (2005: 68) point out, their identities 'can often be expressed through their text-making ... [and] shift in relation to the different ways children feel about the subject of their drawing and writing'. It is children's participa-tion in the visual narrative that plays a part in the 'constuction of the subject' (Chandler 2002: 91) which, as Wright (2010: 36) explains, concerns the child's 'world view and personal identity'.

Children bring their personal and cultural heritages, experiences and identities into their early childhood settings (Brooker 2010), which influence and colour their imaginations and play 'so that children learn to build and connect their funds of knowledge and experience' (Broadhead *et al.* 2010: 181). Privileging democratic learning cultures therefore promotes children's agency, identity and feeling of belonging and is highly relevant for early childhood education. It can inform teachers' and practitioners' understanding – as van Oers (2010: 165) argues:

> An overwhelming mass of studies can be cited to show that schooling is a decisive factor in cognitive development and identity formation, in the distribution of cultural capital and power, as well as in the innovation of culture. It has become clear that the way the teacher organizes classroom activities is crucial for the empowerment of the pupils.

Figure 10.1 Finn: 'I'm three and three quarters; this is how you write three and a half'

Developing agency, developing identities

As Figure 10.1 shows, *children's mathematical graphics* reveal something of their identities, agency and emerging understanding. In the nursery, Finnian (3 years, 9 months) was discussing his age with his peers. The other children in his group were already four and in his estimation were 'bigger'. His mum had told him his current age and Finnian explained, 'I'm *not* three, I'm three and three quarters!' Making some personal letter- or number-like symbols on a whiteboard he explained, 'Look, this is how you write three and three quarters' (top of Figure 10.1).

Drawing on his emotive feelings about his age and perceived position in his group, Finnian used his symbols to communicate his mathematical understanding of fractions, explaining next: 'This is how you write three and a half' (bottom of Figure 10.1). These were Finnian's personal symbols 'for something extremely important to him' and provided a real context for his mathematical development. He was exploring the concept of *nearly* being 4 and attaching symbols to that meaning. The collaborative dialogue with peers and his teacher was vital to his sense of belonging and symbolic understanding, allowing him to assert his agency through his authoritative voice. The discussion continued over several weeks with the children making 'many more examples of their very own fractions', helping Finnian assert his position as a confident and equal member of the group (Carruthers and Worthington 2009a: 24).

Democratic learning cultures

In all the examples in this book, the children's voices combine with those of adults to tell the stories from their perspectives. The examples reveal young children as active, competent, powerful thinkers and meaning-makers, co-constructing meanings and cultures. They show how children use drawing, writing and their *mathematical graphics* to explore anxieties and be in control; to develop, negotiate and justify their sense of belonging (in their family and peer group) and for different communicative purposes including persuasion. They reveal how new media, technologies and popular culture exert influence on both children's feelings and graphicacy. The examples also highlight children's developing understanding of the power of symbolic tools at home and in the nursery.

Understanding appears to emerge through a reciprocal relationship between learners and adults through collaborative dialogue (see Chapter 11). One of the factors we identified was the extent to which adults focused on the children's meanings about their symbolic tools (in play and in their graphics) and how their sensitive dialogue appeared to mediate the children's understanding (Worthington 2010a). Teachers and practitioners often become highly skilled at tuning into children's personal meanings, acknowledging the significance of their feelings and their personal, cultural knowledge. Bringing their 'pedagogy of listening', adults not only focus on the child's embodied meanings but also bring their (adult) cultural insights.

We have argued that it is within democratic learning cultures that individuals can best develop both a strong feeling of agency and a personal identity, and deep understandings of cultural aspects of early childhood education, such as drawing – something that should be within the reach of *all* early childhood settings. However, the children's drawings, writing and *mathematical graphics* included here are not the conventional or 'comfortable' examples that may be expected in school: they are not likely to conform to some adults' expectations and are likely to challenge established perceptions. In doing so they highlight a divide that is often apparent between adults' values concerning children's graphicacy at home or in the nursery and what is expected and accepted in school (Worthington 2010b).

Anning and Ring (2004) have argued that in school children's drawing continues to be poorly understood, but it is likely that children's personal explorations of the symbolic languages of writing and mathematics most often reveal an even greater disjoint in the transition from home to school. Concerns about a narrow 'skills-based' focus on learning for children of 4–5 years have been raised previously (e.g. Adams *et al.* 2004). In school, literacy and mathematics are often seen as separate from children's personal experiences and feelings. Thompson (2003: 188) concludes: 'The way to reintroduce the sovereignty of the learner into the school is to create environments for authentic enquiry . . . Student curiosity itself should be the force that defines the school curriculum' so that they develop 'a strong and positive sense of self-identity' that the curriculum requires (DfES 2007a: 45). Young children need be supported to build personal 'cultural capital' (Bourdieu 1992) and educational cultures in which this is seen as a priority.

Listening to children

Children are born with innate curiosity and a problem-solving aptitude: they want to find out about the world and to make sense of what they see. Their curiosity will be satisfied through their own endeavours and in their own ways, with a supportive listening adult who respects their knowledge and actions. Educational environments that nurture children's natural instinct to enquire will best support children's learning and development. Many mathematicians confirm that mathematics is about reasoning, trying things out and asking 'Why?' (Boaler 2009). Boaler argues that mathematics is also a subject in which people can develop their own ideas, since there are many ways in which mathematical problems can be solved: fortunately young children come to their early years setting with this drive for experimentation and are still asking 'Why?'

In an environment that encourages children's own thinking, socially inclusive relationships are part of the ethos (Lancaster 2003). These relationships mean that the adults understand how to share power so that the children can take part fully in their own play and are able to access opportunities and real choices. Children's ideas and thoughts are central to this environment. Chambers (1994) explains that children need to be active participants in their own learning rather than passive receivers of the education system: they can only be empowered if adults are willing to listen to, and also learn from, them. The environment will be richer as the adult supports the children's adaptations, making additions to the environment. For example, Aimee (4 years, 1 month) took a block from the block area and found that it was the perfect step up to the climbing frame. Charlie (4 years, 10 months) collected all the objects nearby to fill the sandpit and said he was building a fort for his friends to live in. Both Charlie and Aimee have a sense of identity and agency, since they are free to operate in an open play environment.

Teachers talk about children's mathematical play

The nursery headteacher quoted below identified two teachers in her nursery school who really listen and respond to children's mathematical play. The headteacher asked, 'So what makes the difference? What can you tell other teachers about children's mathematical play?'

Emma: Relationships are a massive context. It is a weird relationship not like a traditional 'teachery' relationship but a real relationship.

Carole: So you think that there is mutual respect and you are learning along with the child?

Emma: Almost, you feel that in play any child can come along with anything and that will be fine. The more in-depth your relationship is with the child the more you tune into their play.

Headteacher: What do you mean by 'in depth'?

Emma: It is about knowing the child, their interests and their family. Having seen the child in the home helps, especially for children who do not engage very much. Just knowing the family culture and how they are and their situation. When children play you know their context and it helps with understanding their meaning. Children represent graphicacy in different ways with different motivations and they have shared contexts. Sticky labels become 'supercards' and one child leads this idea and other children buy into this. My children at the moment understand that signs mean power and they are making up their own signs: for example, two crosses mean you can go into the sand and three crosses mean you can go this way and not that way. Once the children explain to the other children they seem to buy into the play. It amazes me the high level of children's play and when you unpick it, it is so complicated. Everything is linked in play, even emotions.

Headteacher: Why do you think the play is so high level here in this nursery?

Emma: The space and freedom.

Carole: It is more the way Emma looks at it. She thinks about it a lot and really engages with the children.

Emma: Children engage in imaginary play and other children's imaginary play. They join in when they want and stop when they want. It is like a train – you get on it and go with it and get off where you like. With children's play it can be a snapshot at that given time but so much comes into that moment. One little opportunity and something sparks it off. Imaginary play is massive and limitless and doesn't need to be confined by government documents. Children in play are exploring huge concepts. Today for example the children were talking down the phone to each other. One of the children was using a diary as a 'booking book' for a campsite. Oliver said, 'I want to stay for two nights.' Isaac said, 'No. I'll put you down for two million nights. Don't worry it's only £1.00 a night.' He then wrote it down in his 'booking' book [see Figures 10.2–10.4]. Isaac had the knowledge and personal experience about campsites and instigated the play and the others went with his idea. This is so different from a themed play area since the children in the campsite play totally owned it and set it up and collected the things they wanted and needed. I got the diaries at our scrapstore and that has just sparked so many ideas for the children's play.

Carole: It is also just taking children's play forward when you hook onto their ideas – for example, when I tied the wool onto the wrists of Yasin because he was in his Spiderman period of play. All the children

Figure 10.2 The children play at bookings for the campsite

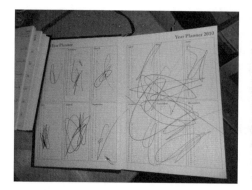

Figure 10.3 Campsite bookings

Figure 10.4 'I'll put you down for two million nights. Don't worry it's only £1.00 a night'

	wanted me to tie wool round their wrists and then everybody was Spiderman and they played with that idea for a long time. It was one of those really wonderful play sessions.
Emma:	Also thinking about the mathematical graphics, as the year goes on and their graphics are encouraged the children really start to understand that they can use this to communicate their mathematical thinking.
Headteacher:	So how do you capture such insightful play observations?
Carole:	Having everything to hand in our 'bum bags'. I keep everything in my bag and my hands are then free to join in the play.
Emma:	We have cameras and notebooks and now I do little film clips.
Carole:	Now I don't even look down when I'm writing, but you have got to be interested. When you take a photograph that usually jogs your memory later on when you are writing up the observations. I think the role of participant observer is crucial.
Emma:	It is important to listen to others about children and read about play. It confirms your thinking and you never know it all.

Both Carole and Emma listen to children's voices as they respond to the their play. Listening to the children they are aware that they learn more about their play and their *mathematical graphics*.

Listening to individual thinking

Children are excited to know about the world; it presents many curiosities to them and they 'zoom in' on aspects that they need further knowledge about. Each child's mind is very individual and if educators truly follow children's own ideas then they will have to follow many different individual paths that children may be following. However, groups of children may have a group cultural interest for a while – for example, one nursery group had an interest in ladders, liked fire engines and enjoyed looking at books on climbing. Their key person took that group to a fire station and while some children kept their interest in ladders, others focused on hoses and bells and going in and out of the fire engine. However, this did not become a topic for all the children, but was part of a whole array of ideas and dialogue the children had initiated and the adult had scaffolded and resourced. The children who were particularly interested in the fire station went to visit it, while another group went to a car manufacturer.

When children are really listened to and have their ideas taken into consideration they have more and more ideas; they brim over with ideas. At one nursery Harry (4 years, 3 months) came in one morning and said, 'I have a thousand ideas!' His key person takes time to listen to the children every morning and this talk time is valued as a crucial learning opportunity in the children's day, ensuring every child is listened to, every day (DCSF 2009a). Boaler strongly advocates that 'Children need to be taught in a culture that values their ability to think and reason, invites their engagement in the breadth of mathematics, and encourages all students to achieve at high levels, even those who struggle for some of the time' (2009: 34).

Inclusion

In a busy play environment it can sometimes be difficult to catch the moments that can be a catalyst for ideas to grow and develop. Are all children heard? Sometimes only the stories and ideas of the most vocal or most able children are heard, since it is easier to listen to those children who are brimming over and are eager to share their thinking. Being aware of gesture and of other languages, disabilities and equal access to play opportunities all have to be considered. Children who are shy and sometimes those whose home culture is radically different from school or nursery can hold back their ideas (Brooker 2010).

Conclusion

All children need to feel comfortable. They need to know they will have space and time to be listened to, and that they will be heard, for only in such cultures will they engage in deep and meaningful mathematical learning.

In the next chapter we will consider aspects of mathematical play, the pedagogy and the play environment.

Reflections

- What opportunities do staff in your setting have to listen to children? Discuss different ways of listening that you have found effective.
- How do children's interests and ideas shape your environment? What impact does this have on their play?
- To what extent do you feel your professional reading supports you in learning more about children? Choosing one or two children's examples in this book, discuss with you colleagues what you have learned about their symbolic play and graphics.

Recommended reading

Dahlberg, G., Moss, P. and Pence, A. (2007) *Beyond Quality in Early Childhood Education and Care: Languages of Evaluation*. London: Routledge.

Drummond, M.J. (2003) *Assessing Children's Learning*. London: David Fulton.

11

Pedagogy that supports mathematical play

In my early teaching years I was in the wrong forest. I paid scant attention to the play and did not hear the stories, though once upon a time, I must have also imagined such wondrous events.

(Paley 1991: 5)

What is this chapter about?

- **A play pedagogy.**
- **Resources and spaces.**
- **Cooking as a medium for mathematical enquiry.**
- **Weaving mathematics throughout the day.**
- **Modelling – a key pedagogical strategy to support children's mathematical thinking.**

This chapter is about the backdrop of play; the panorama that sets the scene for play to take place in the many complex ways it unfolds in the minds of young children. The play environment, both physically and psychologically, is vital to consider as an early years professional and often dictates the quality of the play. We will give examples of environments that provoke mathematical enquiry and build on children's emerging mathematical interests. Mathematics as a holistic experience is discussed with examples of how it can be woven throughout the day.

A play pedagogy

When considering the pedagogy of play there continue to be conflicting ideologies and areas of uncertainty. Stephen (2010) comments that when practitioners were asked about their pedagogy they were uneasy and found it difficult to articulate how they support play learning. The word 'pedagogy' in many practitioners' minds seemed to be linked to direct or didactic teaching, causing them to be hesitant and cautious about how they described their practice (Siraj-Blatchford 1999). Other studies have found that teachers may espouse theories and beliefs in play but in the reality of the classroom these beliefs and theories are not realized (Bennet *et al.* 1997). As play is spontaneous and not pre-planned it is difficult to capitalize on the 'teachable moment' (David and Powell 2007). In their significant study of preschools, Siraj-Blatchford and Sylva (2004) identified 'sustained shared thinking' as a useful pedagogical strategy in the early years setting: this is when the adult shows genuine interest, offers encouragement, clarifies ideas and asks open questions. It extends the children's thinking and helps them to make connections in their learning. Another pedagogical approach is 'scaffolding' (Bruner 1974). The aim here is to support children in their efforts towards the level at which they are capable of working. In this process of gradual support the adult eventually releases the support to the child as they take control and carry on without assistance (Berk and Windsler 1995).

'Co-construction' is a term used in education when the child and adult are seen more as equal partners and there is no predetermined outcome in the teacher's mind (Wood and Attfield 2005). This is about shared meanings, with and the adult and child thinking together, and relies on collaborative dialogue and honest discussions. In the model of co-construction the children have the power and the agency (Konzal 2001): 'Co-construction encourages children and teachers to contribute to developing understandings and is a strategy that develops children's mastery of learning' (Jordan 2003: 51). In the Jordan (2003) study, teachers stated that children were more empowered when interactions were co-constructed in comparison with the outcomes of scaffolded interactions. Co-construction is based on a sociocultural approach to learning and on the premise that 'acting and thinking with others drives learning and at the heart of the process is dialogue' (Stephen 2010).

There is strong evidence that the adult is crucial in a play environment (Broadhead *et al.* 2010). Adults co-constructing understanding with children appears to empower them more than the scaffolding approach, and good teaching is about being flexible enough to do both. Making sensitive judgements about how and when to use these aspects of pedagogy with children is crucial. It is also vital that play-encounters support a 'mutual bridging of meaning' (Stephen 2010).

It is with this sociocultural perspective (Wertsch 1991) in mind that we now describe three possible types of teacher: the insider, the outsider and those who are on the edge of children's learning.

On the inside of children's learning

Being on the inside of children's learning and really knowing the children helps the adult to understand some of the contexts of their talk and play. Paley, a teacher and writer of many early years publications, is an excellent example of an adult who really listens to children and is on the inside of children's learning (Paley 1991). Through observing children, believing in them and carefully considering their thoughts through their stories, Paley has spent a lifetime uncovering their many ideas. Wood puts forward an emergent or responsive pedagogical approach that 'privileges children's cultural practices, meaning and purposes' (2010: 11): when adults are on the inside of children's learning they ask them to clarify their thinking; they speculate with the child. This sort of dialogue provides honest intellectual discussion in which the child is respected

(Bruner 1974). Those adults who are on the inside of children's learning are also supporting them in *thinking* about their learning; in other words, children 'thinking about their thinking' (metacognition). As teachers engage in respectful, real conversations, the children become self-aware and are able to articulate their thoughts. Being on the inside of learning is a valuable way for adults to learn about children's learning: it is a pedagogical tool. Vygotsky (2004) argues that only a deep analysis of play makes it feasible to see its place in young children's learning. Being on the inside of children's learning makes it possible to see the children's own mathematical meanings in their play.

On the outside of children's learning

A numeracy consultant who had always taught older children ('Key Stage 2' in England) was given the remit of supporting Foundation Stage teachers with children's mathematics. With good intentions she showed the teachers how they might set up a role-play area with mathematical content that children could then play in. She decided on the 'Three Bears' theme and set a table with three matching coloured cups, plates, knives and forks. The consultant went to great lengths to find three different-sized teddies, chairs and beds, making matching pillows and covers. She had in her mind a predetermined outcome that the children would go into this role-play area and count the three cups, plates etc. She also thought that as they set the table they would match the colours and take each teddy and match it to the right-sized bed.

The consultant observed the children and was disappointed that they did not do what she'd expected, even when she intervened to show them. The children played with the bears and one child pretended the bear was a baby and took one of the cups to feed the bear, prompting two other children to join in. A fourth child cleared the table, put everything in the cupboard and rearranged the contents several times. The numeracy consultant was completely confused by this and could not understand the children; most of all she could not understand their play. She was on the outside of these young children's learning.

In imaginary play situations children can initiate their own mathematics, but the *outcome* of play cannot be predetermined. Play is unpredictable: it is what children do and it is culturally determined by what the children have seen or know about their real worlds (Brooker 2010). Play is complex and it can be difficult to explain that complexity to adults who are not used to listening to and observing young children play. Bennett *et al.* (1997) argue that there is a great need for adults to understand the meaning of play in children's own terms.

On the edge of children's learning

Quite often you may see, especially in the outside area of a nursery or a school, young children playing and the adults standing around 'supervising' but not observing closely: they are making sure everybody is safe and willingly take the initiative to deal with any first-aid or other needs. Similar to this are those that lean more towards a management role; they keep the time, deal with behaviours and make sure (facilitate) everything is running smoothly. These are worthwhile roles in a play space but if the adults are always on the edge of children's learning they never see or learn about the play, a vital aspect of the early childhood pedagogy. Some adults feel the need to be in a utilitarian role, especially in terms of safety issues, and this often blinds their understanding of children's play (Tovey 2007). These adults are not on the outside of children's play, since they are with children every day, but they are on the edge, looking in and missing opportunities to understand, engage with and extend the children's ideas.

Teachers who practise more adult-directed teaching and leave play on the periphery of the classroom are also on the edge of children's learning. This is because play means that the children have the lead and the adults step into their world and unpick the children's meanings. If teachers are not focusing on children's own play and meanings then they are on the edge of children's learning. Where adults work with children and are on the edge of learning, play is not central to the curriculum they offer to the children with the result that they are unlikely to observe rich play periods and support them.

Environments built on children's emerging ideas

The *Independent Review of Mathematics Teaching in Early Years Settings and Primary Schools* (DCSF 2008a: 34) in England emphasizes the importance of building on children's interest:

> Central to effective mathematical pedagogy in the Early Years is fostering children's natural interest in numeracy, problem solving, reasoning, shapes and measures. Children should be given opportunities in a broad range of contexts, both indoors and outdoors, to explore, enjoy, learn, practise and talk about their developing mathematical understanding. Such experiences develop a child's confidence in tackling problem solving, asking probing questions, and pondering and reasoning answers across their learning.

Yes, but what are children's interests?

There seems to be a misunderstanding about what the term 'children's interests' mean: the term has come into the official documents of the English education system fairly recently and some interpretations are quite broad. One school declared that since the children were interested in dinosaurs they were going to do a whole school topic on them. This drive for everybody doing the same thing appears to haunt some primary schools and it is not uncommon practice that Reception classes in the same school do exactly the same thing for every aspect of their teaching. This is confusing: it raises the question of how, if every child is expected to do the same thing, can children's own ideas be reflected? When children's ideas are truly represented, the graphics displayed will be totally different and the complex individual thinking of the children's ideas will jump out at you. Each child's *mathematical graphics* is different: no two will be the same (Carruthers and Worthington 2006).

When an interest is fermented from a real event

Clare, the practitioner, observed the children's interest as the builders in their nursery built the new extension. The children watched the builders use the spirit levels and Clare bought a range of spirit levels, placing them with the blocks, and the adults supported the children in their exploration of balance and watching the bubble in the spirit level.

One child's interest

Emma, the teacher, purchased four compasses because she had observed Isaac's developing interest in direction and she noticed that in the forest he was using words like 'east' and 'west'. Their collaborative discussions uncovered Isaac's excitement about this area of knowledge.

When listening to a group of children

As a result of children exchanging their ideas and engaging in discussions with their key person in a range of contexts, a 'culture' often develops – a focused interest that many children in the setting become interested in. For example, in a collaborative conversation one key person discovered that the children were interested in ideas about safes and money and how to open a safe. The key person bought a child-sized safe and also some cash boxes with keys. She wrote the number to open the safe on the whiteboard and the children who were interested tried the combination. The staff brought all sorts of foreign money from home and made a display that included three cash registers. The children accessed the display and used the money and cash registers in their play. A child whose first language was Chinese and had been learning English for only four months was focused on the roles of shopkeeper and customer, and exchanging money. She led the play, gesturing that the adult should be the customer and then the shopkeeper. The children chose the resources they needed from the table (see Figure 11.1) to incorporate in their play themes. At the end of the day the table was almost empty since the children had used the resources in a variety of ways in their self-initiated play.

Sustained play theme: children's interest over a period of weeks

Children who are afforded time to play, open resources to experiment with, and adults who listen and are on the inside of children's learning often sustain their play themes over several weeks. This is exemplified in the case study of the television (see Chapter 7). As the children revisit their ideas they build on previous ideas and widen the

Figure 11.1 The interest table as a resource for children

scope of their interests. The remote control described in Chapter 7 became the focus for a while but as the days went on and children returned to this theme, they widened their play to include 'PlayStations' and other channels with different number combinations.

The key roles of participant and non-participant observers

Nearly all the examples and case studies of children's learning in this book include an adult who was carefully interacting with the children and sketching notes at the same time. This model of teaching is vital and a very skilled one. The adult carefully weaves conversation around the children's interests and responds to the child's lead, while also taking observational notes of key moments in the child's actions and talk. Genuine conversations that do not lead the child to an adult-engineered outcome are valuable aspects of uncovering children's own mathematics (Carruthers 2010). Listening, in an active way, without feeling you need to jump in with an immediate response, is another important skill in the role of participant observer. Rather than questioning, when adults comment it can stimulate the child's thinking without the pressure of needing to respond to a question. Questioning can sometimes turn into a 'testing' situation – for example, a favourite question for some adults to ask is 'What colour is this?' Apart from the fact that colours are not mathematical, the question is not genuine because the adult knows the answer. At such times the adult asks this question to test the child's knowledge of colour names: it is a simple style of teaching and does not uncover children's complex thinking.

The participant observer needs to think very carefully about their comments in response to the child. The adult, as participant in the child's enquiries, knows when to step back and senses when *not* to talk. In Chapter 9 the children talk about the number 0. Here, Carole, the teacher, is scaffolding their learning (Bruner 1974), as well as co-constructing with the children their thinking about numerals and taking their learning to the 'zone of proximal development' (Vygotsky 1987). Taking accurate notes to reflect on afterwards helps the teacher understand where a child is in their development and to carefully consider what might consolidate or extend that child's learning. This may then be brought back to the team for discussion.

There is also the role of non-participant observer, who takes notes, observes the children and plans with others to add to the children's experience (see Figure 11.2): this is a less active role but still needs an adult who is skilful in focusing on the

Figure 11.2 Non-participant observer

important aspects of the child's experience (Sheridan 1999). Many good practitioners eventually take part, after they have observed the children's play, because they now know what the play is about and join in, going with the child's lead.

Professional autonomy

All early years professionals need to have professional autonomy to seek out current thinking on play and play pedagogies. Their practice should be informed by what they have observed, and by professional dialogues with colleagues about the personal play theories they are developing. As Wood (2010: 11) states, the 'provision should be informed by knowledge about play and not just by policy-centred versions of education play'. Good early years settings encourage professional dialogue and reflection – for example by providing a professional library.

Procedural and conceptual knowledge

The Researching Effective Pedagogy in the Early Years (REPEY) research (Siraj-Blatchford *et al.* 2002) clearly stated that there is a difference between 'procedural knowledge' which is based on learning facts about the world and how to deal with them, and 'conceptual knowledge' that is concerned with principles and ideas. It is conceptual knowledge, the big ideas, that moves children's thinking on: it is these challenges that children think through when they explore their mathematics through their own *mathematical graphics*. The REPEY project discovered that such learning requires a constructive process that offers the child an opportunity to reflect on and take responsibility for their own thinking and the adult's role is pivotal in providing the 'physical and intellectual environment the child needs' (Siraj-Blatchford *et al.* 2002: 48).

Tuning into and knowing children

The relationship the children have with a key person in the setting is vital to their well being (Elfer *et al.* 2003). Many early childhood settings are busy and active and may seem daunting to a young child. A key person system is where an adult has a small number of children for whom they are responsible on a daily basis. Each child is greeted by that person each morning and goes to that person at various times throughout the day. The system is based on attachment theory (Bowlby 1969) and is a vital relationship in the life of a child. It is this person who knows the child and who is a link to the parents or carers. This is a powerful relationship and one that can hold the knowledge of the child's understanding of mathematics within the setting. Key persons are the teachers and practitioners in early years settings, including schools. A good key person really knows the children and can judge what the individual child needs: they know the children's *rhythm*.

Resources and spaces

It is what a teacher does deliberately, knowing why they are doing it, that makes a difference to a child's mathematical experience (Carruthers and Worthington 2006).

Rich play environments that have the potential for anything do not just 'happen'. As part of excellent pedagogical play strategies, the play environment both inside and outside is key. In England this is one of the key themes of the EYFS curriculum 'enabling environments' (DfES 2007a). Creating and maintaining such an environment takes hard physical work, time and careful planning. If one considers play to be at the heart of the curriculum then planning for it should be at the heart of your strategy. Play resources have different affordances and potential for flexibility, especially where the children have the freedom to make their own novel combinations and transformations (Nutbrown 1994; Carr 2001).

Carefully selected resources are important in the play environment and certain resources particularly promote mathematical opportunities, such as blocks and water. There are also those resources which are deliberately selected with individual children in mind in order to extend their ideas and open up greater play possibilities.

Emma and Karen, teachers in Zurich, Switzerland, comment:

> When we started to think about children's mathematical graphics and how we could support the children in their thinking we decided to combined maths resources and graphic equipment. The children were interested in builders so we set up a construction area and office with dressing-up clothes; inside the drawers we placed tape measures, rulers, calculators, clocks, telephones, clipboards, paper and pencils. I [Emma] also put up a picture of construction workers with blueprints, which provided a model for the children. We discussed the equipment in the area with the children. Then it was very interesting watching them play. They drew plans. The children also referred to numbers on the tape measures, jotting them down to tell 'how big' something was, or 'how much' of something was needed. They measured huge pieces of furniture, carpets etc.

By providing the graphical and mathematical equipment, Karen and Emma supported the children's explorations of their mathematical thinking. The exercise showed them how many of the children already knew a great deal about measurement, and some very big numbers. The teachers said that they felt inspired to continue to develop the children's mathematical understandings.

What to choose

Deliberately planning for the play environment has two levels: the upkeep of general resources that are out in the environment all the time, and the daily changes that are made in responding to children's enquiries. The following examples offer suggestions for resources to enhance the learning of *mathematical graphics* in the play environment.

Resources that support children's understanding of symbolic written languages

At any time a child may choose to write, draw or use graphical implements to represent thinking in any graphical form. Children therefore need opportunities to investigate their enquiries throughout the day in an environment that has readily

Figure 11.3 Children take opportunities in play to draw their narratives on the whiteboards

Figure 11.4 Whiteboard narratives

Figure 11.5 Benjamin (3 years, 7 months) experiments with circular movements on the electronic (interactive) whiteboard

available graphic tools and mathematical equipment. By introducing small white-boards at children's height the teacher and the children can use them for a variety of purposes (see Figures 11.3, 11.4). For example, the boards can be used as message boards for the children and also to model mathematics. The children will constantly use them to explain their thinking and for personal graphics, and they can be very popular, offering an alternative method for children to share their thinking with others. The size of the board means the graphics can be shared and discussed.

Interactive whiteboards (see Figure 11.5) provide a different media that offers collaborative graphical thinking.

Children's own display boards

Both small and large display and clipboards with paper and pencils near to hand are an extremely useful resource, both inside and out (see Figures 11.6–11.10).

Chalks as a medium for graphics outside

Chalks of different lengths and thicknesses provide a great medium for children to create their own symbolic meanings outside. They are non-permanent, can be used on practically anything and can be wiped off easily. For example, Orna (see Figures 11.11–11.13) is using her *mathematical graphics* to write house numbers ('89' and '5') and a note to say 'Don't forget to bring fruit to nursery'. She has seen the teachers write this message on the whiteboard.

Cooking as a medium for mathematical enquiry

Mathematical enquiries are abundant in children's free exploration with a range of ingredients in food sessions (see Figures 11.14, 11.15). For example, 'What speed are

Figure 11.6 Children display their own graphics

Figure 11.7 Children post notes on their board

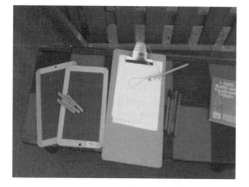

Figure 11.8 Clipboards are always available

Figure 11.9 Numbers on clipboards to encourage thinking about numbers

Figure 11.10 Writing about banana soup

Figure 11.11 Orna scribing on the gazebo (i)

Figure 11.12 Orna scribing on the gazebo (ii)

Figure 11.13 Orna scribing on the gazebo (iii)

Figure 11.14 Using whisks

Figure 11.15 Nathan, Alice and Rhianna making a shopping list: 'olives', 'worms', 'oranges' and 'flour'

we going to set the mixer?', 'How many eggs do we need so that everybody can crack one?', and follow-on enquiries such as 'So one egg has dropped on the floor and we may not have enough, what are we going to do?' (Redcliffe Children 2010).

Number lines

In early childhood settings it is common to see a number line somewhere in the environment. Number lines are essential to provide children with a reference if they want to find a number, and are also beneficial for the teachers and practitioners to use when they want to direct children to a model they can use to help them count. Number lines are most useful if they are used often and when teachers and practitioners talk about them and discuss them with the children. Effective number lines can be easily seen at children's height and it is helpful if there are several number lines, both outside and inside the setting, giving children various models. Number lines that grow with children's interest are more engaging and purposeful (Carruthers 2007b). As children become more proficient with numbers, new numbers can be added or the whole number line changed.

The number sequence does not always need to start with 1 and it intrigues children if you start at 2, or 5, going up beyond 10 or 20. Some children are attracted to larger numbers, and putting numbers in different sequences promotes lots of child-initiated talk in play. For example, when children are interested in numbers and patterns beyond 10, putting '10, 20, 30, 40' or '100', '1000', '10,000', '100,000' and '1,000,000' in their environment supports their enquiry. Some young children really want to know what a million looks like and scribing numbers on a board promotes discussion (see Figure 11.16).

Weaving mathematics throughout the day

Mathematics should not be seen only in discreet adult-directed activities, and practitioners and teachers need to be aware of the mathematics that can happen throughout

Figure 11.16 Number patterns to 1,000,000 in the gazebo

the day and plan for this. Each part of the day is important: for example, snack and lunch time; play time; quiet periods and talk and settling-in periods. Involving children in all the processes of the day makes the mathematics natural and fluent (Carruthers 2010).

Different ways to register

As children settle in, involving them in registration is a meaningful way to start with mathematics. As the year develops, many teachers vary this registration procedure to add variety and also invoke different ways of modelling to represent mathematics. At the beginning of the year children placing their picture on a chart leads to counting how many children are in today. This can be replaced by objects (e.g. a duck or a car) for the children to choose how they want to register. Other alternative strategies include:

- Replacing pictures with names.
- Introducing forms so that children tick or make their own mark to register.
- Making a box of numerals available so that each day a child selects the numeral to match the number of children present, eventually making their own representations.

Macey (Figure 11.17) liked to assume a teacher role and call out names to tick off on a register. Here she created her own list consisting of tallies of various kinds on squared paper explaining, 'The ones going down means they are not allowed in. A cross one means they are allowed in. That means holiday' (pointing to the arc in centre of page).

Ross (Figure 11.18) also made a register: 'That's me,' he said, 'I'm here today.' Looking around to see who else was there he noticed that one of the adults was not in: 'Sue's not here,' he said as he drew a circle, then drawing another circle he added, 'Aiobhinn's not here.'

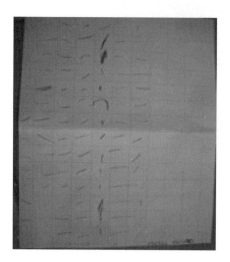

Figure 11.17 Macey's register

Figure 11.18 Ross' register

Talk time

An in-depth study of children's language (Wells 1986) concluded that they need considerable experience of conversation about their own interests and concerns. The routine within one nursery (Carruthers 2008) includes the opportunity for all children to be involved every day in conversation. Children and parents arrive gradually in the morning, as is the case with most early years settings, and this gives the children and parents a chance to talk to the teacher. Resources are laid out at the beginning of the day, sometimes with a mathematical theme, and children can use these as a focus for their talk or play, if they wish. Parents can join in with the children if they choose, and there are always graphical tools readily available.

Real conversation is about exchanging views and opinions, and anybody can start a conversation. Conversation is *not* about putting up hands: there can be silences and pauses for reflection so that conversation is not pressurised. Many of the children's mathematical enquiries start with such a conversation, where there is no abrupt ending and children can go into the next part of the day or stay talking to each other and the teacher. Such talk time at the start of the day gives every child the chance to talk to their key person and their group.

In the example illustrated in Figures 11.19 and 11.20, Hugo, the teacher, had set out tubes and different textures of gravel, salt and sand for the children to explore. Different kinds of graphic tools were also available for the children to use.

In the example shown in Figure 11.21, Amie, the group's key person, set up individual salt trays, containers and spoons for the children to experiment with, and three children chose to spend 45 minutes in dialogue with her about the containers.

Figure 11.19 A variety of resources for talk time

Figure 11.20 Elizabeth and Stacey are using the gravel and carefully putting it through a tube

Figure 11.21 The children are talking about filling, emptying and capacity

Figure 11.22 Stacey lays the table for lunch

Setting tables for lunch

This is a purposeful way to get children involved in everyday, real processes – for example, counting the chairs for each table, putting the knives and forks out, finding enough spoons for every space and counting the number of placemats needed (see Figure 11.22). How many children are having dinner today? Do we have enough yogurts? 'Yasmin does not like yogurt so we only need twenty-two and not twenty-three,' says Matthew. Children have opportunities to count and check the counting and then make decisions about real problems. When you involve children in the everyday processes of life you are respecting them because you are including them, and there is an equal share of power between adult and child.

Snack time

This is similar to setting the table and helping with lunch and provides a purposeful way to involve children. For example, how many milk cartons are needed (see

Figure 11.23 Preparing the milk for snack time

Figure 11.23)? 'Do we need more or less?' 'How many more do we need?' This exercise also involves counting and checking the count, and the children see real mathematical images that they may use in the future in their own *mathematical graphics*.

Modelling – a key pedagogical strategy to support children's mathematical thinking

Teachers modelling different ways to represent mathematics in an informal and relaxed manner when opportunities arise provide children with potential symbolic tools to use when they choose to (Carruthers and Worthington 2006). In *children's mathematical graphics* it is the images they understand that they will be inclined to use. This modelling is not the same as giving children an example after you have taught a mathematics session – the drawback of doing this is that the children invariably copy your model exactly, even if you ask them to choose their own method to use. The power of the teacher's example makes the children think that this is *the* way (Lee 2000) and young children treat examples as something we *expect* them to use even if they do not understand the teacher's method. The most effective support is to model different graphics throughout the week or month (*direct modelling*). When they are participating in children's play, teachers represent mathematical thinking within that context (*indirect adult modelling*). Adults can also support children's images of mathematics in context through signs in the school, nursery or early years setting. When teachers take the register every day they are indirectly modelling a way to gather data and many children, if given the opportunity, make up their own versions of the register (see above and Figures 11.17, 11.18).

Teachers can capitalize on children's own methods and use these as models when discussing different ways to represent. Children themselves will also provide models of mathematics for each other (*peer modelling*) and if they have the right resources and materials at hand in their play they will use their *mathematical graphics* within the play situation.

Through everyday experiences children will have models provided for them, through real contexts outside the nursery or early years setting, in the community (*sociocultural modelling*). Children's home experiences will provide them with images of mathematics through the media and other technologies (*sociocultural modelling*). Their minds are receptive and they store and use all the images that make sense to them.

Using displays

Displays that communicate mathematical symbols will provide an environmental model for children, as shown in Figures 11.24–11.28.

Figure 11.24 Deliberately modelling meaningful mathematics in context through displays

Figure 11.25 Notices provide a focal point with which children can engage

Figure 11.26 Display numerals in various languages (i)

Figure 11.27 Display numerals in various languages (ii)

Figure 11.28 Pigeonholes for messages

Dialogue with parents and carers

Teaching is also about communication with parents about their children. Building a picture of the child's learning with the parent is essential to understand the context of their explorations and parents have a right to be informed about teaching approaches, resources and pedagogical philosophies and models. *Children's mathematical graphics* may not be the conventionally understood way to write mathematics, and parents who have been through the education system and have knowledge of how standardized mathematics should look may at first be unsure of seeing their child's *mathematical graphics* in the same positive light as the child's teacher. *Children's mathematical graphics* may bear little resemblance to the neat row of sums that was the expected notation for mathematics in their school life.

Sharing records of the child's progress and observation of the child is very pleasing to parents and carers and they will be keen to know about their child. Informing parents about the mathematics curriculum at parents' evenings and forums is another way to communicate understanding. Perhaps one of the most successful ways of sharing *mathematical graphics* is putting on an exhibition of the children's graphics, including their *mathematical graphics*: annotated children's graphics on display really help parents understand what children are doing.

Conclusion

This chapter has outlined the key pedagogical strands of *children's mathematical graphics* in play. It has emphasized the teaching points of the environment, the importance of sensitive interactions with children through co-constructing learning and scaffolding, and the understanding that mathematics does not take place only in set adult-directed times but is threaded throughout the day within a culture of mathematical enquiry. The chapter has also detailed the key pedagogical strategy of modelling and given illustrated examples of key teaching experiences. Children can learn mathematics through play provided there is deliberate, thoughtful planning, for and from children's interests, with long uninterrupted periods for them to play. The vital component to this is the teacher, who seeks to understand the children's play and learning.

In the final chapter we consider issues of continuity for mathematics and outline some projects which are supporting important pedagogical development.

Reflections:

- How much time in the day do the children in your setting have for free uninterrupted play?
- Do children have easy access to a variety of graphic equipment throughout the day? For example, pens, paper, chalks, whiteboards, notepads, diaries, clipboards?
- Do practitioners all observe moments of mathematical talk and explorations within the children's play, and use these observations to plan?
- Do practitioners talk with the children about their mathematical ideas, play and explorations?
- Do you provide real mathematical tools and resources such as weighing scales, clocks, calculators and real money?
- What opportunities are there for mathematics in routines?
- Discuss how you might model mathematics to the children.

Recommended reading

Pound, L. (2006) *Supporting Mathematical Development in the Early Years*. Buckingham: Open University Press.

12
Continuity and high expectations

I practiced open-ended questions, the kind that seek no specific answers but rather build a chain of ideas without the need for closure. It was not easy. I felt myself always waiting for the right answer – my answer ... But the goal is the same, no matter what the age of the students; someone must be there to listen, respond, and add a dab of glue to the important words that burst forth. The key is curiosity, and it is curiosity, not answers, that we model. As we seek to learn more about a child, we demonstrate the acts of observing, listening, questioning, and wondering.

(Paley 1986b: 123–7)

In this final chapter we highlight issues concerning continuity across phases for mathematics. We outline two projects from local education authorities that signify important shifts in mathematical teaching and learning, and end with an account of a local Children's Mathematics Network group.

Transitions

Children's interests and strengths do not suddenly change when they move from one setting to another or from one year group to the next, and teachers and practitioners will find it useful to plan together to ensure that children's early fascination with mathematical ideas are celebrated and built upon, to extend their understanding and delight in mathematics and to maintain and strengthen their confidence in mathematical thinking.

The challenges of transitions in England and Scotland have been widely recognized (e.g. Sanders *et al.* 2005; Fabian and Dunlop 2007). A successful start to school is seen as important, 'because of the long-term benefits to learning that this can have, and the need to plan the transition in such a way as to ensure continuity and progression is well documented' (Stephenson and Parsons 2006: 137). The culture of school is different from that of the nursery and children must now become 'pupils'. Stephenson and

Parsons explain that the 'notion that early years education in England covers the age range 3–5 years' (now birth to 5) 'is not helpful . . . Children in Key Stage 1 who are aged 5–7 years are as much "early years" as their peers in Reception classes' (2006: 142).

It has been argued that some of the most significant changes have been found in the transition into Year 1, 'where children often wait to be told what to do next, and have set subjects according to the teacher's agenda' (Fabian and Dunlop 2005: 239). However, many young children in England now often experience a significant change when they leave nursery: in effect, for these children the Foundation Stage ends when they enter Reception. Brooker (2008: 80) refers to Reception classes as a 'half-way house to formal learning, which is intended to make transition more gradual', but goes on to admit that in spite of early years practitioners' 'passion' that:

> Reception children aged 4 to 5 should be exempt from formal instruction and assessment until they have moved into Year 1 . . . [yet] the downward pressure of the National Curriculum, or pressure from parents, obliges them to make increasingly formal provision, especially for literacy and numeracy, long before this.

The outcome is that some English children may begin a 'desk-bound' learning experience from their fourth birthday, implying that as a consequence for such children and parents, 'they are starting school twice over'.

We argue that:

> largely missing from the debate is *curriculum* transition (or *curriculum continuity*) and continuity of the children's experiences of mathematics in the next setting or phase. The evidence is that children experience *problem solving, reasoning and numeracy* in especially contrasting ways, with expectations, opportunities and pedagogy in school that differ from nursery.
>
> (Worthington 2008)

Issues of continuity

Years ago children had opportunities for play throughout their entire primary school education in England, but play has been under considerable threat as concern over 'standards' and 'basic skills' has been enlarged by political and media debates around the world and this has often led to the marginalization of play.

However, observations of children's play often reveal that children far exceed curriculum expectations for mathematics: 'in their play children learn at their highest level' (DfES 2007b). The DCSF booklet *Children Thinking Mathematically* emphasizes that in Reception and Key Stage 1 rich and complex play experiences continue to be important and 'enable children to build on their understanding, making and communicating their meanings in a variety of ways' (DCSF 2009b: 43). The challenge that teachers in Reception and Key Stage 1 face is to understand how to make this a reality for mathematical learning in the classroom (QDCA 2010).

We have shown how 'the natural beginnings of written mathematics start within children's symbolic play . . . More than ever before, children desperately need us to understand how children learn through play and to be advocates for play' (Worthington 2008).

Mathematical continuity

Researchers studying children's experiences in Reception classes began with a view of young children 'as accomplished learners in their own right' (Adams *et al.* 2004: 13). However, their findings revealed an emphasis on 'the smallest building blocks of numeracy ... counting and number recognition' (p. 85). Of greatest concern were 'the limited opportunities for sustained, shared, purposeful talk; for complex, imaginative play; and for authentic, engaging, first-hand experiences' (page 27).

Perhaps it is also the perception of mathematics as a 'hard' subject that is either 'right or wrong' that contributes to the ways in which mathematics is often taught in school, and related to this, young children may be seen to need adults to simplify mathematics since they are young: we caution against such a view since this can only lead to viewing 4-year-olds as 'blank slates' and beginning with the sort of mathematical activities that Adams *et al.* (2004) describe. Yet, as many of the examples in this book show, this does not need to be so. Continuity of approaches so that the child's experiences of mathematics grow from what they have previously known can ensure they continue to reach their potential and is especially important where observations of play can often reveal that children exceed curriculum expectations.

However, while the Foundation Stage emphasizes that practitioners 'build on what children already know and understand' (QCA 2000: 11), it seems clear that teachers have been uncertain what this means for mathematics and that 'official' guidance has previously been sorely lacking on this important aspect of children's mathematical development and the pedagogy to support it. In the light of this significant omission, it seems to us especially hard that official reports such as Ofsted's annual reports of mathematics continually raise concerns about children's understanding in mathematics, including their 'written' mathematics and calculations (and by association, what teachers do).

Children's own *mathematical graphics* have an important bridging function between concrete and mental operations, and between internal mental representations and their external 'written' mathematics (Carruthers and Worthington 2008). And as we have shown, their own graphics enable children to bridge the divide between their early informal marks and the standard (abstract) written language of mathematics. In one local authority where all early years teachers and practitioners had support in developing mathematics in their settings, the early years adviser raised a concern: 'I've just been to one of the schools and seen some fabulous stuff – lovely subtraction with numbers to 20 from a 4-year-old. The teacher is worrying about transition to Y1 now where they might be back to subtracting within 10 on a worksheet . . .'

In addition to acknowledging the importance of *children's mathematical graphics*, the two-year *Independent Review of Mathematics Teaching* makes specific recommendations for those working in early childhood education (DCSF 2008a). The report's authors acknowledge that transition 'directly affects the young learner in mathematics' and requires 'full account to be taken of the child's accomplishments', arguing that 'Familiar approaches to children should [also] be maintained in Year 1'. The report encourages teachers in Year 1 'to increase opportunities for active, independent learning through play, as in the EYFS, to ensure a continuation of positive attitudes to mathematics' (pages 40, 41). The *Review* reiterates that there 'is then a need for a coherent approach overall to the progression from EYFS to Year 1, and it is essential that the momentum in learning in mathematics is maintained through this transition. This makes it all the more important that more attention is given to this question of continuity' (page 63). *Children's mathematical graphics* throughout the Foundation Stage and Key Stage 1 will ensure that they develop a good understanding of the written language of mathematics. It will also ensure *continuity* in written mathematics as children move from nursery to the Reception class and later into Key Stage 1, and 'requires full account to be taken of the child's accomplishments, and needs to reflect the perspectives of a range of contributors, especially parents' (DCSF 2008a: 40).

Children using and applying mathematics

It could be argued that children solving problems is at the heart of mathematics and indeed we have emphasized the importance of children's deep-level thinking as they use their *mathematical graphics*. The *Independent Review of Mathematics Teaching* (2008) argues that relatively few children attain Point 8 – 'uses developing mathematical ideas and methods to solve practical problems' – in any of the three mathematical assessment scales. This is the aspect of mathematics that in Key Stage 1 is known as 'using and applying mathematics' and refers to children solving problems, representing, enquiring, reasoning and communicating mathematics (DfES 2006). The DCSF suggests that evidence for this 'is often missed – not because children lack competence in this area, but because the opportunity for them to demonstrate their developing thinking skills is not there'. When children are encouraged 'to take charge of their learning and to think for themselves', observations can reveal 'the sophistication of young children's capacity for solving problems' (DCSF 2010: 30).

Local authority advisers leading change: evidence-based, grass-roots research

Lesley, a local authority adviser, who leads this project in the south-east of England, explains:

> For the last three years our focus was on raising standards in early years mathematics; every setting and school has been involved through courses and meetings and visits. Each year we targeted a group of Reception classes. This was within the local authority's larger remit of improving pedagogy and achievement in mathematics. Children's mathematical graphics is seen as a contributory part of the whole mathematics focus in the authority. Increasingly teachers shared their examples of children's mathematical graphics.
>
> At one school the children are playing with two-digit numbers, discussing thousands. They say they have never done this before. Five-year-olds playing with millions, exploring number lines, dividing odd numbers three ways – with all the evidence. I have one teacher who is particularly excited – she says she's intrigued to see how everything leads into standard calculations. When we started a couple of years ago I hadn't anticipated that I'd be looking so much at the

environment and while we found that we were a very good at print-rich, literary environment, as often as not there were no numbers and if there were they usually stopped at nine. Looking at the learning environment led naturally into writing resources, visual numbers and hence into our expectations of early written maths.

The teachers and practitioners expressed that the difference was the paper in the environment and the staff looking at the graphics alongside talking and thinking. Previously with maths if they had a gifted child they would just have gone 'schooly' with maths from a workbook. Now that the children are such independent thinkers, the adults are just really a resource. Throughout the authority the expectation is much higher in mathematics for children.

On analysing the data at the end of the first year of this initiative Lesley said there was all-round improvement, especially in scale Point 8. This is highly significant, since throughout the country the scores for this scale point are lower than the other Foundation Stage Profile (FSP) points. FSP Point 8 is arguably the most instrumental scale point as it lays the basis for all mathematics.

There is no doubt that the success of this project can be attributed to Lesley's drive and enthusiasm, with the encouragement and support of the early years team. Throughout the past three years we have been delighted to hear of the impact that this initiative has had and believe that its success points to similar potential for other local authorities.

The Ethnic Minority and Travellers' Achievement Service (EMTAS) project

Joy, the local authority adviser who leads this project in a London Borough explains:

For many years I have had a professional interest in early years development and was keen to steer a project that would make a difference for the youngest learners within the local authority and in particular those at risk of underachieving – namely, certain black African and black Caribbean and advanced bilingual boys – since they had been identified as underachieving. Mindful of the Williams Review [DCSF 2008a] and widening gaps in attainment, I knew this area would provoke debate and keen interest in Brent schools.

I wanted to organize an action research project to investigate whether or not this specific intervention of children's mathematical graphics would make a positive impact on these learners as well as other pupils in the classroom. For example, a teacher's observation of Daniel (4 years, 8 months) and Susan (4 years, 10 months), revealed the following.

'Daniel and Susan stack some of the blocks on top of each other next to the measuring stick. They measure 18 cms. Daniel writes 18 on his clipboard. I watch him and Susan, they take the blocks of wood off the measuring stick and then measure them one by one – Daniel records the height of each block and Susan measures it. When they have finished they show me what they have done. It looks like 18, 15, 15, 15. However, I ask them to tell me about their representations and they reply, "We used a 1, 5, 1, 5, 1, 5, and they made 18." They have calculated $1 + 5 + 1 + 5 + 1 + 5 = 18$.'

These and many more examples were a watershed moment for many of the teachers involved in this action research. Now, their standpoint had changed, each could see the potential to exploit mathematics in their practice: they would seek out, encourage, support, listen to, extend and celebrate children's mathematical experiences. Young children's mathematical ability is astounding and challenging their teaching professionals, causing them to re-evaluate and have even higher expectations. I just can't wait to repeat this study again next year.

Kylie, one of the project's teachers added:

As a teacher, children's mathematical graphics have given me a much deeper understanding of my class's knowledge and understanding of number, calculations and space, shape and measures and I have become a lot more confident in my assessment. As adults we use a lot of shared thinking to make sure we never miss an opportunity to encourage children to use graphics and to offer real situations to use them.

This is another successful project which, though with a different remit had a similar outcome, and again we can attribute its success to the organization and management of the leaders and the enthusiasm of the teachers involved. This project has implications for the teaching and learning of mathematics for those groups that are seen to be at risk of underachieving.

Local Children's Mathematics Network groups

In 2003 we established the 'Children's Mathematics Network', an international, not-for-profit organization for teachers, practitioners, students, researchers and teacher educators working with children in the birth to 8 age range. It is a grass-roots network, with children and teachers at its heart. We had both been leading face-to-face professional development courses for many years but knew that these could only ever provide an introduction. The Children's Mathematics Network encourages teachers to take ownership of their own continuing professional development by organizing their own local *mathematical graphics* interest groups.

This model of professional development has been highly successful. For example, the Bristol Children's Mathematics Network group was the focus of research by independent researchers from the University of Oxford, the University of Bristol, King's College London and the University of Birmingham from the Researching Effective CPD in Mathematics Education (RECME) project. RECME is the largest UK research project into professional development in mathematics education to date. This research focused on a range of development initiatives for early years, primary and secondary mathematics teaching. The RECME research findings (NCETM 2009) confirmed our own expectations about the potential of empowering teachers in their mathematics teaching (Carruthers and Worthington 2009b), and the importance of trusting teachers and practitioners and giving them 'ownership' of their professional development.

The final report of the RECME research explains that 'The overarching aim of the research was to investigate the interrelated factors that contribute to "effective" CPD for teachers of mathematics and the outcomes of the study will inform future CPD and impact on policy-making in education' (NCETM 2009). The final report includes six case studies to highlight various positive aspects and the Children's Mathematical Network professional development initiative is one of these. Among the key findings was:

> a suggestion that teachers who gave accounts of changes in practice that they reported as sustained and profound had often been involved in their CPD in focusing on student learning and reflecting on the relationship between student learning and their own professional practice in the classroom. We also found that some of those teachers who were beginning to develop as leaders of CPD themselves were involved in reading and reflecting on educational research findings and it is this aspect on which this account focuses.
>
> (Back and Joubert 2009: 13)

The researchers noted that this initiative focused 'on careful consideration and analysis of children's mathematics and the ways in which professionals can support and encourage the children's mathematical thinking' (NCETM 2009: 65). The report observes that 'The standard of the mathematical understanding, thinking and reasoning that the displays revealed was far higher than the specified curriculum objectives for children of this age' (page 64).

Starting a local Children's Mathematics Network group may be a possible way forward for teachers who are beginning to develop their understanding of children's mathematics.

Conclusion

This book has been written for teachers to answer their enquiries about *children's mathematical graphics* and play. It celebrates young children's creativity and the many perceptive teachers who have shared their perspectives and told their children's stories in this book. We believe that insightful and sensitive teachers are the key to enhancing children's play worlds. What makes the difference is those teachers who understand the deep level thinking that play evokes, that will be the grounding to solving increasing complex algorithms and problems as the children move through school.

For most of our career we were busy classroom teachers trying in hurried moments, to understand the layers of complexities in children's play that unfolded each day. As we shared with each other our experiences about children's thinking we resolved that there is always more to know and understand, and we continue to seek to understand. In Spike Milligan's words.

> I am of the opinion that children are not just small homo sapiens – they are an entirely different species, with a secret world that only very perceptive adults have any real knowledge of. I have. Lucky me.

(Milligan 2004)

We believe that in the pages of this book the children's and their teachers' voices shine through.

References

Adams, S., Alexander, S., Drummond, M.J. and Moyles, J. (2004) *Inside the Foundation Stage: Re-creating the Reception Year*. London: Association of Teachers and Lecturers (ATL).

Agar, S. (n.d.) www.omniglot.com, accessed 15 May 2009.

Aldrich, F.K. and Sheppard, L. (2000) Graphicacy: the fourth 'r'?, *Primary Science Review*, 64, 8–11, September–October.

Alland, A. Jr (1983) *Playing with Form: Children Draw in Six Cultures*. New York: Columbia University Press.

Anning, A. (2003) Pathways to the graphicacy club: the crossroads of home and pre-school, *Journal of Early Childhood Literacy*, 3(1): 5–31.

Anning, A. (2004) The role of drawing in young children's journeys towards literacy, *Education*, 3–13, June, 2000, 32–8.

Anning, A. and Ring, K. (2004) *Making Sense of Children's Drawings*. Maidenhead: Open University Press.

Askew, M., Brown, M., Rhodes, V., Wiliam, D. and Johnson, D. (1997) *Effective Teachers of Numeracy: Report of a Study Carried out for the Teacher Training Agency*. London: King's College, University of London.

Athey, C. (2007) *Extending Thought in Young Children: A Parent-teacher Partnership*, 2nd edn. London: Paul Chapman.

Aubrey, C (1997) *Mathematics Teaching in the Early Years: An Investigation of Teachers' Subject Knowledge*. London: The Falmer Press.

Back, J. and Joubert, M. (2009) Reflecting on practice in early years' settings: developing teachers' understandings of children's early mathematics, *Proceedings of the British Society for Research into Learning Mathematics*, 29(1): 13–18.

Bakhtin, M.M. (1986) *Speech Genres and Other Late Essays*. Austin, TX: University of Texas Press.

Bennett, N., Wood, E. and Rogers, S. (1997) *Teaching Through Play: Teachers' Thinking and Classroom Oractice*. Buckingham: Open University Press.

Berk, L. and Winsler, A. (1995) *Scaffolding Children's Learning: Vygotsky and Early Childhood Education*. Washington, DC: National Association for the Education of Young Children.

Blatchford, P., Hallam, S., Ireson, J., Kutnick, O. and Creech, A. (2008) *Classes, Groups and Transitions: Structures for Teaching and Learning*, research survey 9/2: the Primary Review. Cambridge: University of Cambridge.

Boaler, J. (2009) *The Elephant in the Classroom: Helping Children Learn to Love Maths*. London: Souvenir Press.

Bowlby, J. (1969) *Attachment and Loss*, vol. 1. London: Hogarth Press.

Bourdieu, P. (1977) *Outline of a Theory of Practice*. Cambridge: Cambridge University Press.

Bourdieu, P. (1992) *Language and Symbolic Power*. Oxford: Basil Blackwell.

Brizuela, B. (2004) *Mathematical Development in Young Children: Exploring Notations*. New York: Teachers College Press.

Brizuela, B. (2006) Young children's notations for fractions, *Educational Studies in Mathematics*, 62(3): 281–305.

Broadhead, P. (2004) *Early Years Play and Learning: Developing Social Skills and Cooperation*. London: RoutledgeFalmer.

Broadhead, P., Wood, E. and Howard, J. (2010) Understanding playful learning and pedagogies – towards a new research agenda, in P. Broadhead, J. Howard. and E. Wood (eds) *Play and Learning in Educational Settings*. London: Sage.

Brody, H. (1988) *Maps and Dreams*. Vancouver, British Colombia: Douglas and McIntyre.

Brook, T. (2009) *Vermeer's Hat: The Seventeenth Century and the Dawn of the Global World*. London: Profile Books.

Brooker, L. (2010) Learning to play in a cultural context, in P. Broadhead, J. Howard and E. Woods (eds) *Play and Learning in the Early Years*. London: Sage.

Bruce, T. (2004) *Developing Learning in Early Childhood*. London: Sage.

Bruner, J.S. (1974) Child's play, *New Scientist*, 62: 126–8.

Burt, C. (1921) *Mental and Scholastic Tests*. London: P.S. King & Son.

Caillois, R. (2001) *Man, Play and Games*. Urbana, IL: University of Illinois Press.

Carr, M. (2001) *Assessment in Early Childhood Settings*. London: Paul Chapman Publishing.

Carraher, T., Carraher, D. and Schliemann, A. (1985) Mathematics in the streets and in school, *British Journal of Developmental Psychology*, 3: 21–9.

Carruthers, E. (1997a) *Number: A Developmental Theory – A Case Study of a Child from Twenty to Forty-four Months*, unpublished M. Ed. dissertation, University of Plymouth.

Carruthers, E. (1997b) Talking numbers: a developmental link between literacy and numeracy, *Early Education*, Summer: 5–6, 6–19.

Carruthers, E. (2007a) Under threes thinking mathematically? Paper presented at the European Early Childhood Education Research Association Conference, Prague. 29 August–1 September 2007.

Carruthers, E. (2007b) A number line in the nursery classroom: a vehicle for understanding children's number knowledge, *Early Years*, 18(1), autumn: 9–14.

Carruthers, E. (2008) The importance of young children's mark-making: beginnings, context, meanings in mathematics, *The Foundation Stage Forum*, December.

Carruthers, E. (2010) How children use graphics to support mathematical thinking, *Early Years Educator*, June: 39–44.

Carruthers, E. and Worthington, M. (2005) Making sense of mathematical graphics: the development of understanding abstract symbolism, *European Early Childhood Education Research Association Journal*, 13(1): 57–79.

Carruthers, E. and Worthington, M. (2006) *Children's Mathematics: Making Marks, Making Meaning*, 2nd edn. London: Sage.

Carruthers, E. and Worthington, M. (2008) Children's mathematical graphics: young children calculating for meaning, in I. Thompson, (ed.) *Teaching and Learning Early Number*, 2nd edn. Maidenhead: Open University Press.

Carruthers, E. and Worthington, M. (2009a) Marking time, *Nursery World*, 1 October: 24–5.

Carruthers, E. and Worthington, M. (2009b) An early years CPD initiative for mathematics: the power of collaborative, 'grassroots' learning, *Proceedings of the British Society for Research into Learning Mathematics*, 29(1), March: 25–30.

Chambers, P. (1994) The origins and practice of participatory rural appraisal, *World Development*, 22(7): 953–69.

Chandler, D. (2002) *Semiotics: The Basics*. London: Routledge.

Chatwin, B. (1987) *The Songlines*. London: Pan.

Clay, M. (1975) *What Did I Write?* London: Heinemann.

Cobb, P., Yackel, E. and Wood, T. (1992) A constructivist alternative to the representational view of mind in mathematics education, *Journal for Research in Mathematics Education*, 23: 2–33.

Collins, D. (1976) *The Human Revolution: From Ape to Artist*. London: Book Club Associates/Phaidon Press.

Cook, V. (2001) *Second Language Learning and Language Teaching*. London: Arnold.

Coté, J.E. and Levine, C.G. (2002) *Identity Formation, Agency and Culture: A Scoal Pshychological Synthesis*. Mahwah, NJ: Lawerence Erlbaum.

Cox, M. (2005) *The Pictorial World of the Child*. Cambridge: Cambridge University Press.

Cox, M., Koyasu, M. and Hiranuma, H. (2001) Children's human figure drawing in the UK and Japan: the effects of age, sex and culture, *British Journal of Developmental Psychology*, 19: 275–92.

Dahlberg, G., Moss, P. and Pence, A. (2007) *Beyond Quality in Early Childhood Education and Care: Languages of Evaluation*. London: Routledge.

David, T. and Powell, S. (2007) Beginning at the Beginning, in J. Moyles (ed.) *Beginning Teaching, Beginning Learning*, 3rd edn. Maidenhead: Open University Press.

David, T., Gooch, K., Powell, S. and Abbott, L. (2002) *Birth to Three Matters*. London: Sure Start.

De Ruyter, D. and Conroy, J. (2002) The formation of identity: the importance of ideas, *Oxford Review of Education*, 28(4): 509–22.

DCSF (Department for Children, Schools and Families) (2008a) *Independent Review of Mathematics Teaching in Early Years Settings and Primary Schools*. Final Report – Sir Peter Williams, June 2008. London: DCSF.

DCSF (Department for Children, Schools and Families) (2008b) *Mark Making Matters*. London: DCSF.

DCSF (Department for Children, Schools and Families) (2009a) *Independent Review of the Primary Curriculum: Final Report*. London: DCSF.

DCSF (Department for Children, Schools and Families) (2009b) *Children Thinking Mathematically: PSRN Essential Knowledge for Early Years Practitioners*. London: DCSF.

DCSF (Department for Children, Schools and Families) (2010) *Finding and Exploring Young Children's Fascinations – Strengthening the Quality of Gifted and Talented Provision in the Early Years*, http://nationalstrategies.standards.dcsf.gov.uk/node/348649.

DfES (Department for Education and Skills) (2004) *Every Child Matters: Change for Children*. London: DfES.

DfES (Department for Education and Skills) (2006) *Using and Applying Mathematics*. London: DfES, http://nationalstrategies.standards.dcsf.gov.uk/node/47324?uc=force_uj, accessed 10 June 2009.

DfES (Department for Education and Skills) (2007a) *Practice Guidance for the Early Years Foundation Stage*. London: DfES.

DfES (Department for Education and Skills) (2007b) *Card 4.1: Learning and Development: Play and Exploration*. London: DfES.

DiSessa, A., Hammer, D., Sherin, B. and Kolpakowski, T. (1991) Inventing graphing: meta-representational expertise in children, *Journal of Mathematics Behaviour*, 10: 117–60.

Dowker, A. (2004) *What Works for Children with Mathematical Difficulties?* Research report no. 554. London: DfES.

Drury, R. (2007) *Young Bilingual Learners at Home and at School.* Stoke-on-Trent: Trentham Books.

Dyson, A.H. (1997) *Writing Superheroes: Contemporary Childhood, Popular Culture, and Classroom Literacy.* New York: Teachers College Press.

Dyson, A.H. (2001) Where are the childhoods in childhood literacy? An exploration in outer (school) space, *Journal of Early Childhood Literacy*, 1: 9–39.

Edmiston, B. (2008) *Forming Ethical Identities in Early Childhood Play.* London: Routledge.

Einarsdottir, J., Dockett, S. and Perry, B. (2009) Making meaning: children's perspectives expressed through drawings, *Early Childhood Development and Care*, 179(2): 217–32.

Elfer, P., Goldschmeid, E. and Selleck, D. (2003) *Key Persons in the Nursery: Building Relationships for Quality Provision.* Abingdon: David Fulton.

Ernest, P. (1991) *The Philosophy of Mathematics Education.* London: Falmer Press.

Ernest, P. (2006) A semiotic perspective of mathematical activity: the case of number, *Educational Studies in Mathematics*, February: 409–10.

Fabian, H. and Dunlop, A.W. (2005) The importance of play in the transition to school, in J. Moyles, (ed.) *The Excellence of Play.* Maidenhead: Open University Press.

Fabian, H. and Dunlop, A.W. (2007) The first days at school, in J. Moyles (ed.) *Beginning Teaching, Beginning Learning.* Maidenhead: Open University Press.

Ferreiro, E. and Teberosky, A. (1982) *Literacy Before Schooling.* Portsmouth, NH: Heinemann.

Finkel, I. and Seymour, M. (2008) *Babylon: City of Wonders.* London: The British Museum.

Flewitt, R. (2005) Is every child's voice heard? Researching the different ways 3-year-old children communicate and make meaning at home and in a pre-school playgroup, *Early Years*, 25(3): 207–22.

Fuson, K.C., Wearne, D., Hiebert, J.C., Murray, H.G., Human, P.G., Olivier, A.I., Carpenter, T.P. and Fennema, E. (1997) Children's conceptual structures for multidigit numbers and methods of multidigit addition and subtraction, *Journal for Research in Mathematics Education*, 28(2): 130–62.

Gallon, R.M. (1981) Presidential address, *The National Society for Art Education Journal*, 8(1): 3.

Gell, A. (1998) *Art and Agency: Towards a New Anthropological Theory.* Oxford: Clarendon Press.

Gibson, J.J. (1979) *The Ecological Approach to Visual Perception.* Boston, MA: Houghton Mifflin.

Gifford, S. (1990) Young children's representations of number operations, *Mathematics Teaching*, 132: 64–71.

Gifford, S. (2005) *Teaching Mathematics 3–5.* Maidenhead: Open University Press.

Ginsburg, H. (1977) Learning to count: computing with written numbers: mistakes, in H. Ginsburg, *Children's Arithmetic: How They Learn It and How You Teach It.* New York: Van Nostrand.

Golumb, C. (2002) *Child Art in Context: A Cultural and Comparative Perspective.* Washington, DC: American Psychological Association.

Gombrich, E.H. (1960) *Art and Illusion: A Study in the Psychology of Pictorial Representation.* Oxford: Phaidon Press.

Goodenough, F. (1926) *Measurement of Intelligence by Drawings.* New York: Harcourt, Brace & World.

Gopnik, A., Melzoff, A. and Khul, S. (1999) *How Babies Think: The Science of Childhood.* London: Weidenfeld & Nicholson.

Goswami, U. (2008) *Cognitive Development: The Learning Brain.* Hove: Psychology Press.

Green, J. (2009) *Between the Earth and the Air: Multimodality in Arandic Sand Stories*, Ph.D. dissertation, Melbourne: University of Melbourne.

Gregory, E. and Williams, A. (2000) *City Literacies: Learning to Read Across Generations and Cultures.* London: Routledge.

Gura, P. (1992) *Exploring Learning: Young Children and Blockplay*. London: Paul Chapman.

Hall, E. (2007) Mixed messages: the role and value of drawing in early education, paper presented at the British Educational Research Association (BERA) annual conference, Institute of Education, University of London, 5–8 September.

Hall, E. (2010) Identity and young children's drawings: power, agancy, control and transformation, in P. Broadhead, J. Howard and E. Wood (eds) *Play and Learning in the Early Years*. London: Sage.

Harrison, K.D. (2007) *When Languages Die*. Oxford: Oxford University Press.

Hill, R. (1996) *Windows on the Mind*, BBC Open University programme, June.

Holland, D., Lachicotte, W., Skinner, D. and Cain, C. (1998) *Identity and Agency in Cultural Worlds*. London: Routledge.

Holland, P. (2003) *We Don't Play with Guns Here*. Maidenhead: Open University Press.

Hughes, M. (1986) *Children and Number: Difficulties in Learning Mathematics*. Oxford: Blackwell.

Huizinga, J. (1950) *Homo Ludens: A Study of the Play-element in Culture*. Boston, MA: The Beacon Press.

Isaacs, S. (1929) *The Nursery Years*. London: Routledge & Kegan Paul.

Janzen, M. (2008) Where is the (postmodern) child in early childhood education research? *Early Years*, 28(3): 287–98.

John-Steiner, V. (1985) The road to competence in an alien land: a Vygotskiian perspective on bilingualism, in J. Wertsch (ed.) *Culture, Communication and Cognition: Vygotskiian Perspectives*. Cambridge: Cambridge University Press.

Jordan, B. (2003) *Professional Development Making a Difference for Children: Co-constructing Understandings in Early Childhood Centres*, unpublished Ph.D. thesis, Massey University, New Zealand.

Juster, N. (2008) *The Phantom Tollbooth*. London: HarperCollins.

Karmiloff-Smith, A. (1992) *Beyond Modularity: A Developmental Perspective on Cognitive Science*. Cambridge, MA: MIT Press.

Kelly, C. (2010) *Hidden Worlds: Young Children Learning Literacy in Multicultural Contexts*. Stoke-on-Trent: Trentham Books.

Kenner, C. (2004) *Becoming Biliterate: Young Children Learning Different Writing Systems*. Stoke-on-Trent: Trentham Books.

Konzal, J.L. (2001) Collaborative enquiry: A means of creating a learning community, *Early Childhood Research Quarterly*, 16: 95–115.

Kress, G. (1993) Against arbitrariness: the social production of the sign as a foundational issue in critical discourse analysis, *Discourse and Society*, 4(2): 169–91.

Kress, G. (1997) *Before Writing: Rethinking the Paths to Literacy*. London: Routledge.

Kress, G. (2003) *Literacy in the New Media Age*. London: Routledge.

Kress, G. and van Leeuwen, T. (2001) *Multimodal Discourse: The Modes and Media of Contemporary Communication*. London: Arnold.

Lancaster, L. (2001) Staring at the page: the function of gaze in a young child's interpretation of symbolic forms, *Journal of Childhood Literacy*, 1(2): 131–52.

Lancaster, L. (2003) Moving into literacy: how it all begins, in N. Hall, J. Larson and J. Marsh (eds) *Handbook of Early Childhood Literacy*. London: Sage.

Lancaster, L. (2007) Representing the ways of the world: how children under three start to use syntax in graphic signs, *Journal of Early Childhood Literacy*, 7(2): 123–54.

Lancaster, P. and Broadbent, V. (2003) *Listening to Young Children*. Maidenhead: Open University Press.

Lawrence, V., Luck, D. and Stevenson, C. (2008) *The Baby Room Project*. Northampton: Northamptonshire Local Authority.

Lee, C. (2000) Modelling in the mathematics classroom, *Mathematics Teaching*, June: 28–31.

Lerman, S. (1989) Constructivism, mathematics and mathematics education, *Educational Studies in Mathematics*, 20: 211–23.

Light, P. (1985) The development of view specific representation considered from a socio-cognitive standpoint, in N.H. Freeman and M. Cox (eds) *Visual Order: The Nature and Development of Pictorial Representation*. Cambridge: Cambridge University Press.

Lowenfeld, V. and Brittain, W. (1987) *Creative and Mental Growth*. New York: Macmillan.

Luquet, G. (2001) *Children's Drawings ('Le Dessin Enfantin')*. London: Free Association Books.

Luria, A.R. (1998) The development of writing in the child, in M.K. Oliviera and V. Valsiner (eds) *Literacy in Human Development*. Stamford, CT: Ablex.

Malchiodi, C. (1998) *Understanding Children's Drawings*. New York: The Guilford Press.

Marsh, J. (2004) The techno-literacy practices of young children, *Journal of Early Childhood Research*, 2(1): 51–66.

Marsh, J. (ed.) (2005a) *Popular Culture, New Media and Digital Literacy in Early Childhood*. London: RoutledgeFalmer.

Marsh, J. (2005b) Ritual, performance and identity construction, in J. Marsh (ed.) *Popular Culture, New Media and Digital Literacy in Early Childhood*. London: RoutledgeFalmer.

Marsh, J., Brooks, G., Hughes, J., Ritchie, L., Roberts, S. and Wright, K. (2005) *Digital Beginnings: Young Children's Use of Popular Culture, Media and New Technologies*. Sheffield: University of Sheffield Literacy Research Centre.

Marshall, S. (1968) *An Experiment in Education*. Cambridge: Cambridge University Press.

Matthews, J. (1998) The representation of events and objects in the drawings of young children from Singapore and London: implications for the curriculum, *Early Years*, 19(1): 90–109.

Matthews, J. (1999) *The Art of Childhood and Adolescence: The Construction of Meaning*. London: Falmer Press.

Matthews, J. (2003) *Drawing and Painting: Children and Visual Representation*. London: Paul Chapman.

McMahon, P., Nyheim, T. and Schwartz, A. (2006) After the tsunami: lessons form re-construction, *The McKinsey Quarterly*, 1: 95–105, www.ieco.clarin.com/2008/05/12/afts06.pdf, accessed 10 February 2010.

Mehler, J. and Dupoux, E. (1994) *What Infants Know*. Oxford: Blackwell.

Mercer, N. (2000) *Words and Minds: How We Use Language to Think Together*. London: Routledge.

Millard, E. (2006) Transformative pedagogy: teachers creating a pedagogy of fusion, in K. Pahl and J. Rowsell (eds) *Travel Notes from the New Literacy Studies*. Clevedon: Multilingual Matters.

Milligan, S. (2004) *A Children's Treasury of Milligan*. London: Virgin Books.

Moll, L., Amanti, C., Neff, D. and Gonzalez, N. (1992) Funds of knowledge for teaching: using a qualitative approach to connect homes and classrooms, *Theory into Practice*, 31(2), Spring: 132–41.

Moss, P. (2007) Bringing politics into the nursery: early childhood education as a democratic practice, *European Early Childhood Education Research Journal*, 15(1): 5–20.

Moyles, (1989) *Just Playing? The Role and Status of Play in Early Childhood Education*. Buckingham: Open University Press.

Moyles, J. (1994) *The Excellence of Play*. Buckingham: Open University Press.

Moyles, J. (2010) Foreword, in P. Broadhead, L. Wood. and J. Howard (eds) *Play and Learning in Educational Settings*. London: Sage.

Munn, P. (1994) The early development of literacy and numeracy skills, *European Early Childhood Education Research Journal*, 2(1): 5–18.

Nash, D. (1998) Ethnocartography: understanding central Australian geographic literacy, paper presented to the Australian Anthropological Society annual conference, Canberra, 2 October, www.anu.edu.au/linguistics/nash/abstracts/cartog.html, accessed 19 May 2009.

NCETM (National Centre for Excellence in Teaching Mathematics) (2009) *Final Report: Researching Effective CPD in Mathematics Education.* London: DCSF.

Newman, J. (1984) *The Craft of Children's Writing.* New York: Scholastic Book Services.

Nunes, T. and Bryant, P. (1996) *Children doing Mathematics.* Oxford: Blackwell.

Nunes, T., Schliemann, A. and Carraher, D. (1993) *Street Mathematics and School Mathematics.* Cambridge: Cambridge University Press.

Nutbrown, C. (1994) *Threads of Thinking: Young Children Learning and the Role of Early Education.* London: Paul Chapman.

Nutbrown, C. (2001) Wide eyes and open minds: assessing and respecting children's early achievements, in J. Collings and D. Cook (eds) *Understanding Learning, Influences and Outcomes.* London: Paul Chapman.

Ofsted (Office for Standards in Education) (2002) Ofsted Invitation Conference for Primary Teachers: a conference designed to share best practice in problem solving, communication and reasoning in primary mathematics, Brunel University, March.

Ofsted (Office for Standards in Education) (2008) *Mathematics: Understanding the Score,* www.ofsted.gov.uk/Ofsted-home/Publications-and-research/Browse-all-by/Documents-by-type/Thematic-reports/Mathematics-understandin g-the-score, accessed 12 January 2010.

Oliveira, Z.M.R. and Valsiner, J. (1997) Play and imagination: the psychological construction of novelty, in A. Fogle, M. Lyra and J. Valsiner (eds) *Dynamics and Indeterminism in Developmental and Social Processes.* Hillsdale, NJ: Erlbaum.

Olson, D. (1994) *The World on Paper: The Conceptual and Cognitive Implications of Writing and Reading.* Cambridge: Cambridge University Press.

Pahl, K. (1999a) *Transformations: Meaning Making in the Nursery.* Stoke-on-Trent: Trentham Books.

Pahl, K. (1999b) Making models as a communicate practice – observing meaning making in a nursery, *Reading,* 33(3): 114–19.

Pahl, K. (2001) Texts as artefacts crossing sites: map making at home and at school, *Reading,* 35(3): 120–5.

Pahl, K. (2002) Ephemera, mess and miscellaneous piles: texts and practices in families, *Journal of Early Childhood Literacy,* 2(2): 145–66.

Pahl, K. and Rowsell, J. (2005) *Literacy and Education: Understanding the New Literacy Studies in the Classroom.* London: Paul Chapman.

Pahl, K. and Rowsell, J. (eds) (2006) *Travel Notes from the New Literacy Studies.* Clevedon: Multilingual Matters.

Paley, V.G. (1984) *Boys and Girls: Superheros in the Doll Corner.* Chicago, IL: University of Chicago Press.

Paley, V.G. (1986a) *Mollie is Three.* Chicago: University of Chicago Press.

Paley, V.G. (1986b) On listening to what the children say, *Harvard Educational Review,* 56(2): 122–32.

Paley, V.G. (1991) *The Boy Who Would Be a Helicopter: The Uses of Storytelling in the Classroom.* Cambridge, MA: Harvard University Press.

Paley, V.G. (2004) *A Child's Work: The Importance of Fantasy Play.* Chicago: University of Chicago Press.

Pape, S.J. and Tchoshanov, M.A. (2001) The role of representation(s) in developing mathematical understanding, *Theory into Practice,* 40(2): 118–25.

Papert, S. (1993) *Mindstorms: Children, Computers and Powerful Ideas.* London: Harvester Wheatsheaf.

Pellegrini, A.D. and Smith. P.K. (2005) *The Nature of Play: Great Apes and Humans*. London: The Guilford Press.

Peterson, P., Fennema, E., Carpenter, T. and Loef, M. (1989) Teachers pedagogical content beliefs in mathematics, *Cognition and Instruction*, 6(1): 1–40.

Piaget, J. (1958) *The Child's Construction of Reality*. London: Routledge & Kegan Paul.

Pimm, D. (1995) *Symbols and Meanings in School Mathematics*. London: Routledge.

Poland, M. and van Oers, B. (2007) Effects of schematising on mathematical development, *European Early Childhood Education Research Journal*, 15(2): 269–93.

Porascy, J., Young, E. and Patton, J. (1999) The emergence of graphicacy, *The Journal of General Education*, 48(2): 103–10.

Potts, R. (1996) *Humanity's Descent*. New York: William Morrow.

Pound, L. (2006) *Supporting Mathematical Development in the Early Years*, 2nd edn. Maidenhead: Open University Press.

Pramling, N. (2009) External representation and the architecture of music: children inventing and speaking about notations, *British Journal of Music Education*, 26(3): 273–91.

QCA (Qualifications and Curriculum Authority) (1999) *Teaching Written Calculations*. London: QCA.

QCA (Qualifications and Curriculum Authority) (2000) *Curriculum Guidance for the Foundation Stage*. London: QCA.

QCA (Qualifications and Curriculum Authority) (2003) *Foundation Stage Profile*. London: QCA.

QDCA (Qualifications and Curriculum Development Agency) (2010) *The First Year of the Early Years Foundation Stage*, report no. QDCA/10/4783. London: Qualifications and Curriculum Council.

Redcliffe Children (2010) *Children's Food Based Creative Thinking*. Bristol: Redcliffe Children's Centre.

Ring, K. (2001) Young children drawing: the significance of the context, paper presented at the British Educational Research Association (BERA) annual conference, University of Leeds, September.

Ring, K. (2005) Supporting young children drawing: developing a role, paper presented at the EECERA 15th annual conference, 'Young Children as Citizens: Identity, Belonging, Participation', Dublin, September.

Riojas-Cortez, M. (2001) Preschoolers' funds of knowledge displayed through sociodramatic play episodes in a bilingual classroom, *Early Childhood Education Journal*, 29(1): 35–40.

Robinson, A. (1995) *The Story of Writing*. London: Thames & Hudson.

Rogoff, B. (2003) *The Cultural Nature of Human Development*. Oxford: Oxford University Press.

Rousseau, C., Lacroix, L., Singh, A., Gauthier, M.F. and Benoit, M. (2005) Creative expression workshops in school: prevention programs for immigrant and refugee children, *Canadian Child Adolescence Psychiatric Review*, 14(3): 77–80.

Sanders, D., White, G., Burge, B., Sharp, C., Eames, A., McEune, R. and Grayson, H. (2005) *A Study of Transition from the Foundation Stage to Key Stage 1*, NFER research report. Slough: National Foundation for Educational Research.

Pulley Sayre, A. & Sayre. J. (2004) *One is a Snail, Ten is a Crab*. London: Walker Books.

Seeger, (1998) Representations in the mathematics classroom: re-elections and constructions, in F. Seeger, J. Voigt and V. Werschescio (eds) *The Culture of the Mathematics Classroom*. Cambridge: Cambridge University Press.

Selter, C. (1998). Building on children's mathematics: a teaching experiment in grade three, *Educational Studies in Mathematics*, 36: 1–27.

Sheridan, M.D. (revised and updated by J. Harding and L. Meldon-Smith) (1999) *Play in Early Childhood: From Birth to Six Years*, 2nd edn. London: Routledge.

Siraj-Blatchford, I. (1999) Early childhood pedagogy: practice principles and research, in P. Mortimer (ed.) *Understanding Pedagogy and its Impact on Learning*. London: Paul Chapman.

Siraj-Blatchford, I. and Sylva, K. (2004) Researching pedagogy in English pre-schools, *British Educational Research Journal*, 30(5): 713–30.

Siraj-Blatchford, I., Sylva, K., Muttock, S., Gilden, R. and Bell, D. (2002) *Researching Effective Pedagogy in the Early Years*. London: DfES.

Starkey, P. and Cooper, R.G. Jr (1980) Perception of number by human infants, *Science*, 210: 1033–5.

Stephen, C. (2010) Pedagogy: the silent partner in early years learning, *Early Years: An International Journal of Research and Development*, 30: 15–29.

Stokrocki, M. (1994) Through Navajo children's eyes: cultural influences on the representational abilities, *Visual Anthropology*, 7: 27–67.

Simpson, J. (2006) Sand talk and how to record it, in *Transient Languages and Cultures*, http://blogs.usyd.edu.au/elac/2006/10/sand_talk_and_how_to_record_it.html, accessed 17 May 2009.

Stephenson, M. and Parsons, M. (2006) Expectations: effects of curriculum change as viewed by children, parents and practitioners, in A.-W. Dunlop and H. Fabian (eds) *Informing Transitions in the Early Years: Research, Policy and Practice*. Maidenhead: Open University Press.

Sully, J. (2000) *Studies of Childhood*. London: Free Association Books.

Sulzby, E. (1985) Kindergarteners as writers and readers, in M. Farr (ed.) *Advances in Writing Research, Vol. 1: Children's Early Writing Development*. Norwood, NJ: Ablex.

Sutton-Smith, B. (1997) *The Ambiguity of Play*. Cambridge, MA: Harvard University Press.

Sylva, K. and Wiltshire, J. (1993) The impact of early education on children's later development: a review prepared for the RSA inquiry 'Start Right'. *European Early Childhood Education Research Journal*, 1(1): 17–40.

Sylva, K., Roy, C. and Painter, M. (1980) *Childwatching at Playgroup and Nursery School*. London: Grant McIntyre.

Terwel, J., Van Oers, B., Van Dijk, I. and Van den Eeden, P. (2009) Are representations to be provided or generated in primary mathematics education? *Educational Research and Evaluation*, 15(1): 25–44.

Thiel, O. (2010) Teachers' attitudes towards mathematics in early childhood education, *European Early Childhood Education Research Journal*, 18(1): 105–15.

Thompson, C.M. (1999) Action, autobiography and aesthetics in young children's self-initiated drawings, *The International Journal of Art and Design Education*, 18(2): 155–61.

Thompson, D. (2003) Early childhood literacy education, wakefulness and the arts, in L. Bresler and M. Thompson (eds) *The Arts in Children's Lives: Context, Culture and Curriculum*. London: Kluwer.

Tizard, B. and Hughes, M. (1984) *Young Children Learning: Talking and Thinking at Home and at School*. London: Fontana.

Tomasello, M. (1999) *The Cultural Origins of Human Cognition*. Cambridge, MA: Harvard University Press.

Tovey, H. (2007) *Playing Outdoors: Spaces and Places, Risk and Challenge*. Maidenhead: Open University Press.

UNICEF (2009) *Convention on the Rights of the Child*. Office of the High Commissioner for Human Rights. http://www2.ohchr.org/english/law/crc.htm, accessed 14 December 2009.

Van Leeuwen, T. (2005) *Introducing Social Semiotics*. London: Routledge.

Van Oers, B. (2000) The appropriation of mathematical symbols: a psychosemiotic approach to mathematics learning, in P. Cobb., E. Yackel and K. McClain (eds) *Symbolizing and Communicating in Mathematics Classrooms*. Mahwah, NJ: Lawrence Erlbaum.

Van Oers, B. (2001) Educational forms of initiation in mathematical culture, *Educational Studies in Mathematics*, 46: 59–85.

Van Oers, B. (2002) Teachers' epistemology and the monitoring of mathematical thinking in early years classrooms, *European Early Childhood Education Research Journal*, 10(2): 19–30.

Van Oers, B. (2005) The potentials of imagination, *Inquiry: Critical Thinking Across the Disciplines*, 24(4): 5–17.

Van Oers, B. (2010) Teaching as a collaborative activity: an activity theoretical contribution to the innovation of teaching, *Mind, Culture and Activity*, 12(2): 165–7.

Van Oers, B. and Poland, M. (2007) Schematising activities as a means for encouraging young children to think abstractly, *Mathematics Education Research Journal*, 19(2): 10–22.

Vellom, R.P. and Pape, S.J. (2000) EarthVision 2000: examining students' representations of complex data, *School Science and Mathematics*, 100: 426–39.

Vile, A. (1999) What can semiotics for mathematics education? *Research in Mathematics Education*, 1(1): 87–102.

Vygotsky, L.S. (1930) The problem of the cultural development of the child, in L.S. Vygotsky and A. Luria, *Studies of the History of Behaviour: Ape, Primitive Man, and Child: Essays in the History of Behaviour*. Hillsdale, NJ: Lawrence Erlbaum Associates.

Vygotsky, L.S. (1978) *Mind and Society: The Development of Higher Mental Processes*. Cambridge, MA: Harvard University Press.

Vygotsky, L.S. (1983) The prehistory of written language, in M. Martlew (ed.) *The Psychology of Written Language*. Chichester: Wiley.

Vygotsky, L.S. (1987) The development of imagination in childhood, in R.W. Rieber and A.S. Carton (eds) *The Collected Works of L.S. Vygotsky*, vol. 2. New York: Plenum Press.

Vygotsky, L.S. (2004) Imagination and creativity in childhood, *Journal of Russian and East European Psychology*, 42(1): 7–97.

Wade, D., Mackenzie, I. and Kennedy, S. (1995) *Nomads of the Dawn: The Penan of the Borneo Rain Forest*. San Francisco: Pomegranate Books.

Wells, G. (1986) *The Meaning Makers: Children Learning Language and Using Language to Learn*. Portsmouth, NH: Heinemann.

Wertsch, J. (1991) *Voices of the Mind: A Sociocultural Approach to Mediated Action*. Cambridge, MA: Harvard University Press.

Wilmot, P.D. (1999) Graphicacy as a form of communication, *South African Geographical Journal*, 81(2): 91–5.

Wilson, B. (2002) Becoming Japanese: *Manga*, children's drawings and the construction of national character, in L. Bresler and C.M. Thompson (eds) *The Arts in Children's Lives: Context, Culture and Curriculum*. London: Kluwer.

Wood, E. (2008) Conceptualising a pedagogy of play: international perspectives from theory, policy and practice, in D. Kuschner (ed.) From Children to Red Hatters®, diverse images and issues of play, *Play and Culture Studies*, 8: 168–89. Lanham, MD: University Press of America.

Wood. E. (2009) Developing a pedagogy of play, in A. Anning, J. Cullen and M. Fleer (eds) (2009) *Early Childhood Education: Society and Culture*. London: Sage.

Wood, E. (2010) Developing integrated pedagogical approaches to play and learning, in P. Broadhead, L. Wood. and J. Howard (eds) *Play and Learning in Educational Settings*. London: Sage.

Wood, E. and Attfield, J. (1996) *Play, Learning and the Early Childhood Curriculum*. London: Paul Chapman.

Wood, E. and Attfield, J. (2005) *Play, Learning and the Early Childhood Curriculum*, 2nd edn. London: Paul Chapman.

Worthington. M. (2008) Playful pedagogy and transitions in children's mathematics, *The Foundation Stage Forum*, December.

Worthington, M. (2009) Fish in the water of culture: signs and symbols in young children's drawing, *Psychology of Education Review*, 33(1): 37–46.

Worthington, M. (2010a) Play as a complex landscape: imagination and symbolic meanings, in P. Broadhead, L. Wood. and J. Howard (eds) *Play and Learning in Educational Settings*. London: Sage.

Worthington, M. (2010b) Imagination and graphicacy: young children as powerful agents of their own learning, paper presented at Anglia Ruskin University conference, 'Early Childhood Curriculum: Policy and Pedagogy in the 21st Century: An International Debate', Chelmsford, April.

Worthington, M. (2010c) 'This is a *different* calculator – with computer games on': reflecting on children's symbolic play in the digital age, in J. Moyles (ed.) *Thinking About Play: Developing a Reflective Approach*. Maidenhead: Open University Press.

Worthington (2010e) Symbolic play and the emergence of graphicacy in early childhood mathematics: a semiotic perspective, submitted to *Research in Mathematics Education*, summer 2010.

Worthington, M. and Carruthers, E. (2003) *Children's Mathematics: Making Marks, Making Meaning*. London: Paul Chapman.

Wright, S. (2010) *Understanding Creativity in Early Childhood*. London: Sage.

Index

Locators shown in *italics* refer to figures.